PLEASE STAMP DATE DUE, BOTH BELOW AND ON CARD

D1737333

WITHDRAWN
Geology
QE522 .V65 1999
Volcanoes in the Quaternary
CALTECH LIBRARY SERVICES

Volcanoes in the Quaternary

Geological Society Special Publications
Series Editors
A.J. Fleet
R. E. Holdsworth
A. C. Morton
M. S. Stoker

GEOLOGICAL SOCIETY SPECIAL PUBLICATION NO. 161

Volcanoes in the Quaternary

EDITED BY

C. R. FIRTH
Department of Geography & Earth Sciences, Brunel University, UK

and

W. J. McGUIRE
Benfield Greig Hazard Research Centre, Department of Geological Sciences,
University College London, UK

1999
Published by the Geological Society
London

THE GEOLOGICAL SOCIETY

The Geological Society of London was founded in 1807 and is the oldest geological society in the world. It received its Royal Charter in 1825 for the purpose of 'investigating the mineral structure of the Earth' and is now Britain's national society for geology.

Both a learned society and a professional body, the Geological Society is recognized by the Department of Trade and Industry (DTI) as the chartering authority for geoscience, able to award Chartered Geologist status upon appropriately qualified Fellows. The Society has a membership of 8600, of whom about 1500 live outside the UK.

Fellowship of the Society is open to persons holding a recognized honours degree in geology or a cognate subject and who have at least two years' relevant postgraduate experience, or not less than six years' relevant experience in geology or a cognate subject. A Fellow with a minimum of five years' relevant postgraduate experience in the practice of geology may apply for chartered status. Successful applicants are entitled to use the designatory postnominal CGeol (Chartered Geologist). Fellows of the Society may use the letters FGS. Other grades of membership are available to members not yet qualifying for Fellowship.

The Society has its own Publishing House based in Bath, UK. It produces the Society's international journals, books and maps, and is the European distributor for publications of the American Association of Petroleum Geologists (AAPG), the Society for Sedimentary Geology (SEPM) and the Geological Society of America (GSA). Members of the Society can buy books at considerable discounts. The Publishing House has an online bookshop (*http://bookshop.geolsoc.org.uk*).

Further information on Society membership may be obtained from the Membership Services Manager, The Geological Society, Burlington House, Piccadilly, London W1V 0JU (Email: *enquiries@geolsoc. org.uk;* tel: +44 (0)171 434 9944).

The Society's Web Site can be found at *http://www.geolsoc.org.uk/*.The Society is a Registered Charity, number 210161.

Published by The Geological Society from:
The Geological Society Publishing House
Unit 7, Brassmill Enterprise Centre
Brassmill Lane
Bath BA1 3JN, UK

(*Orders*: Tel. +44 (0)1225 445046
Fax +44 (0)1225 442836)
Online bookshop: *http://bookshop.geolsoc.org.uk*

First published 1999

The publishers make no representation, express or implied, with regard to the accuracy of the information contained in this book and cannot accept any legal responsibility for any errors or omissions that may be made.

The Geological Society of London 1999. All rights reserved. No reproduction, copy or transmission of this publication may be made without written permission. No paragraph of this publication may be reproduced, copied or transmitted save with the provisions of the Copyright Licensing Agency, 90 Tottenham Court Road, London W1P 9HE. Users registered with the Copyright Clearance Center, 27 Congress Street, Salem, MA 01970, USA: the item-fee code for this publication is 0305-8719/99/$15.00.

British Library Cataloguing in Publication Data
A catalogue record for this book is available from the British Library.

ISBN 1-86239-049-5
ISSN 0305-8719

Typeset by Bath Typesetting, Bath, England

Printed by Alden Press, Osney Mead, Oxford OX2 0EF, UK

Distributors

USA
  AAPG Bookstore
  PO Box 979
  Tulsa
  OK 74101-0979
  USA
  Orders: Tel. + 1 918 584-2555
        Fax +1 918 560-2652
        Email bookstore@aapg.org

Australia
  Australian Mineral Foundation Bookshop
  63 Conyngham Street
  Glenside
  South Australia 5065
  Australia
  Orders: Tel. +61 88 379-0444
        Fax +61 88 379-4634
        Email bookshop@amf.com.au

India
  Affiliated East-West Press PVT Ltd
  G-1/16 Ansari Road, Daryaganj,
  New Delhi 110 002
  India
  Orders: Tel. +91 11 327-9113
        Fax +91 11 326-0538

Japan
  Kanda Book Trading Co.
  Cityhouse Tama 204
  Tsurumaki 1-3-10
  Tama-shi
  Tokyo 206-0034
  Japan
  Orders: Tel. +81 (0)423 57-7650
        Fax +81 (0)423 57-7651

Contents

Preface

New Zealand, North Island volcanic province

CASSIDY, J., LOCKE, C. A., MILLER, C. A. & ROUT, D. J.:   1
The Auckland volcanic field, New Zealand: geophysical evidence for its eruption history

GILES, T. M., NEWNHAM, R. M., LOWE, D. J. & MUNRO, A. J.:   11
Impact of tephra fall and environmental change: a 1000 year record from Matakana Island, Bay of Plenty, North Island, New Zealand

NEWNHAM, R. M., LOWE, D. J. & ALLOWAY, B. V.:   27
Volcanic hazards in Auckland, New Zealand: a preliminary assessment of the threat posed by central North Island silicic volcanism based on the Quaternary tephrostratigraphical record

East African Rift Valley and the Mediterranean

SCOTT, S. C. & SKILLING, I. P.:   47
The role of tephrachronology in recognizing synchronous caldera-forming events at the Quaternary volcanoes Longonot and Suswa, south Kenya Rift

HARDIMAN, J. C.:   69
Deep sea tephra from Nisyros Island, eastern Aegean Sea, Greece

VINCIGUERRA, S., GAROZZA, S., MONTALTO, A. & PATANÈ, G.:   89
Eruptive and seismic activity at Etna Volcano (Italy) between 1977 and 1991

Late Quaternary eruptions in Iceland

GRATTON, J., GILBERTSON, D. & CHARMAN, D.:   109
Modelling the impact of Icelandic volcanic eruptions upon the prehistoric societies and environment of northern and western Britain

GONZALEZ, S., JONES, J. M. & WILLIAMS, D. L.:   125
Characterization of tephras using magnetic properties: an example from SE Iceland

CHARMAN, D. J. & GRATTON, J.:   147
An assessment of discriminant function analysis in the identification and correlation of distal Icelandic tephras in the British Isles

GRATTON, J. & SADLER, J.: 161
Regional warming of the lower atmosphere in the wake of volcanic
eruptions: the role of the Laki fissure eruption in the hot summer of 1783

BRAYSHAY, M. & GRATTAN, J.: 173
Environmental and social responses in Europe to the 1783 eruption of
the Laki fissure volcano in Iceland: a consideration of contemporary
documentary evidence

Hazard assessment

CHESTER, D. K., DIBBEN, C., COUTINHO, R., DUNCAN, A. M., 189
COLE, P. D., GUEST, J. E. & BAXTER, P. J.:
Human adjustments and social vulnerability to volcanic hazards: the case
of Furnas Volcano, São Miguel, Açores

SOLANA, M. C. & APARICIO, A.: 204
Reconstruction of the 1706 Montaña Negra eruption. Emergency
procedures for Garachico and El Tanque, Tenerife, Canary Islands

Index 217

Preface

It has long been suggested that there is a causal link between volcanic activity and Quaternary environmental changes. The impact of individual eruptions on regional and global weather conditions was first identified by Franklin (1789) and has been assessed in detail by Lamb (1972). A more contentious issue is the relationship between volcanic activity and Quaternary glaciations. Indeed, some workers (e.g. Humphreys 1940; Wexler 1952) suggested that volcanic activity may have driven the larger scale Quaternary glacial fluctuations. More recently, however, there has been a growing body of evidence which suggests that the converse view is true, namely that Quaternary environmental changes resulted in increased volcanic activity (Zielinski et al. 1996; McGuire et al. 1997). In addition, Quaternary scientists have used tephra layers as chronological horizons and have recently started to suggest that volcanic events may produce not only short-term changes in climate but also variation in regional vegetation patterns and in the distribution of society. It thus appears that a full understanding of the eruptive histories of volcanoes and Quaternary environmental change requires co-operation and collaboration between these two fields of science. It is hoped that this volume provides the first step in this process.

The seed of this volume was planted during a European Communities funded research project looking at the link between volcanic activity and Quaternary sea-level change. The project illustrated that much was to be gained from the interaction of Quaternary scientists and volcanologists. As a consequence a conference was jointly convened by the Volcanic Studies group of the Geological Society and the Quaternary Research Association, the principal aim being to ensure that scientists from the two disciplines could see how their research interacted. The following collection of papers reflects the diversity of research being conducted in this field, ranging from methodological papers which illustrate how eruptive chronologies can be determined, through studies which look at the regional impacts of eruptions, to assessments of modern volcanic hazards.

This volume contains a collection of 13 papers which, together, form a representative cross-section of research into Quaternary volcanic activity and associated environmental impact. The papers have been grouped on a geographical basis and attempt to illustrate the diverse focus of research in each area. The first three papers are associated with the volcanic province of New Zealand's North Island. These provide an assessment of its eruptive history, the impact of eruptive events on local vegetation and an evaluation of the volcanic hazard based on tephrostratigraphic records. The next 3 papers relate to the East African Rift Valley and the Mediterranean, outlining the importance of tephrostratigraphic records in the determination of eruptive chronologies and the link between eruptive and seismic activity on one major volcanic edifice. The following 5 papers deal with Late Quaternary eruptions in Iceland, in particular how such events can be identified at local and distal sites via analysis of tephras and what impacts such events can have on regional weather conditions and society. The final 2 papers are based around research conducted on Atlantic volcanic islands and provide detailed study of hazard assessment in such areas.

I would like to thank the referees who provided valuable comments on the papers accepted for publication and the support provided by my co-author Bill McGuire. I would also like to thank the staff of the Geological Society Publishing House and Burlington House for their patience and determination which ensured that this volume was published.

Callum Firth
West London

References

FRANKLIN, B. 1789. Meteorological imaginations and conjectures. *Memoirs Manchester Literary and Philosophical Society*, **2**, 373–377.

HUMPHREYS, W. J. 1940. *Physics of the air*. New York, MacGraw Hill.

LAMB, H. H. 1972. *Climate past, present and future* (Volume 1). Methuen, London.

MCGUIRE, W. J., HOWARTH, R. J., FIRTH C. R. *et al.* 1997. Correlation between rate of sea-level change and frequency of explosive volcanism in the Mediterranean. *Nature*, **389**, 473–476.

WEXLER, H. 1952. Volcanoes and climate. *Scientific American*, **186**, 74–80.

ZIELINSKI, G. A., MAYEWSKI, P. A., MEEKER, L. D., WHITLOW, S. & TWICKLER, M. S. 1996. An 110,000-year record of explosive volcanism from the GISP2 (Greenland) ice core. *Quaternary Research*, **43**, 109–118.

References to this volume

It is recommended that reference to all or part of this book should be made in one of the following ways:

FIRTH, C. R. & MCGUIRE, W. J. (eds) 1999. *Volcanoes in the Quaternary*. Geological Society, London, Special Publication, **161**.

GILES, T. M., NEWNHAM, R. M., LOWE, D. J. & MUNRO, A. J. 1999. Impact of tephra fall and environmental change: a 1000 year record from Matakana Island, Bay of Plenty, North Island, New Zealand. *In*: FIRTH, C. R. & MCGUIRE, W. J. (eds) *Volcanoes in the Quaternary*. Geological Society, London, Special Publication, **161**, 11–26.

The Auckland volcanic field, New Zealand: geophysical evidence for its eruption history

JOHN CASSIDY[1], CORINNE A. LOCKE[1], CRAIG A. MILLER[1] & DAVID J. ROUT[2]

[1] *Department of Geology, The University of Auckland, Private Bag 92019, Auckland, New Zealand*
[2] *Department of Geology and Geophysics, The University of Western Australia, Nedlands, Perth, WA 6907, Australia*

Abstract: The Late Quaternary monogenetic basalt volcanoes of the Auckland volcanic field exhibit styles of eruption ranging from phreatomagmatic to magmatic. New detailed aeromagnetic and other geophysical data from the southern half of the field provide constraints on the style and relative timing of eruptions. Concealed basalt bodies are shown to be common beneath maars indicating deep excavation by phreatomagmatic events and subsequent filling by magma. Depth of excavation is unrelated to the presence of surficial and potentially saturated Plio-Pleistocene sediments but commonly involves magma–water interaction in widespread aquifers within the underlying Miocene sediments. Coincident anomalous bulk magnetization directions show that at least three volcanoes were active contemporaneously and suggest a very short duration of activity which is probably typical of other centres in the field. These results emphasize the spasmodic nature of activity in the field over the last 140 ka culminating in the most recent centre, Rangitoto, which erupted after a long quiescent period and represents a large spasm of activity, confined to a single centre. The locations of contemporaneous centres are tentatively correlated with the regional NNW–SSE structural trend.

The Auckland volcanic field is the youngest of several predominantly basaltic Pliocene to Recent intraplate volcanic associations in the northern North Island, New Zealand (Heming 1980; Smith 1989), comprising about 50 eruption centres within an area of 360 km^2 (Fig. 1). The field has been active for the last 140 ka (Wood 1991) within which time it has produced a total of about 4 km^3 of eruptive materials (Smith & Allen 1993). The Auckland volcanoes are monogenetic, each the result of a single-eruption sequence (Kermode 1992); there is no evidence for distinct periods of activity separated by long time intervals at any one centre. Many monogenetic basalt fields, e.g. the East Eifel field, Germany (Schmincke *et al.* 1983) and the Pinacante field, Mexico (Gutmann 1979) exhibit eruptive mechanisms ranging from magmatic (Strombolian/Hawaiian scoria cones and lava flows) to phreatomagmatic (maars and tuff rings). The Auckland volcanic field shows all gradations between these two end-member eruption styles (Houghton *et al.* 1991).

This paper presents data from the southern half of the field, covering 21 of the volcanic centres. The results of aeromagnetic, palaeomagnetic and gravity studies are used to deduce the extent of phreatomagmatic activity and the relative timing of

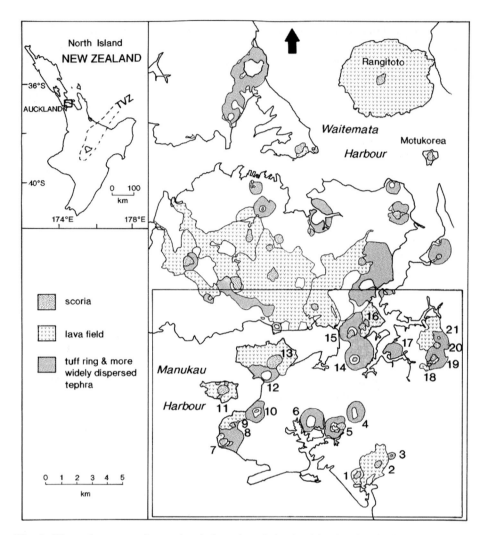

Fig. 1. The volcanoes and associated deposits of the Auckland volcanic field (after Allen 1992). The inset shows the location of the field with respect to North Island, New Zealand, and to the Taupo Volcanic Zone (TVZ). The lower box within the main figure defines the area of Fig. 2. Volcanic centres in the southern half of the field, surveyed in this study, are 1: McLaughlins Hill (Matakarua); 2: Wiri (Manurewa); 3: Ash Hill; 4: Kohoura; 5: Crater Hill; 6: Pukaki Crater; 7: Maungataketake; 8: Otuataua; 9: Pukeiti; 10: Waitomokia; 11: Puketutu; 12: Mangere Lagoon; 13: Mt Mangere; 14: Robertson Hill; 15: Mt Richmond; 16: McLennan Hills; 17: Pukekiwiriki; 18: Hampton Park; 19: Otara Hill; 20: Green Hill; 21: Styaks Swamp.

eruptions in this part of the field. These results also have implications for hazard assessment in the Auckland volcanic field.

Geological setting and volcanic hazard

The Auckland volcanic field occurs in an area of possible intraplate extension (Smith 1989), some 400 km west of the currently active plate boundary which lies east of New Zealand. The geology of the Auckland region is dominated by Miocene sediments, consisting of alternating mudstone–sandstone sequences, which are underlain at an unknown depth (though probably about 1–2 kms: Kermode 1992) by Mesozoic metasedimentary basement rocks. These basement rocks are upthrown and outcrop about 10 km to the east of the area shown in Fig. 1 and exhibit Late Tertiary–Quaternary block faulting with NNW–SSE and orthogonal WSW–ENE orientations. The NNW–SSE orientation is also the dominant structural trend throughout northernmost New Zealand i.e. almost orthogonal to the axis of arc volcanism (TVZ in Fig. 1) through central North Island (Sporli 1989). There is very little evidence for Late Quaternary faulting in the Auckland region (Sporli 1989) and no evidence for recent faulting, though very low level microseismicity does occur in the upstanding basement to the east (Cassidy et al. 1986).

The erupted rocks of the Auckland volcanic field are dominantly alkali basalt or basanite with less common tholeiite, transitional basalt and nephelinite (Heming & Barnet 1986). In the southern half of the field the volcanoes were erupted onto thin cover rocks consisting of sandy Plio-Pleistocene sediments which include some pumiceous deposits. Significant aquifers occur in both the Plio-Pleistocene and the Miocene sediments (Waterhouse 1967; Rout 1991), the latter containing significant grit and conglomerate horizons.

Age information for the Auckland volcanoes is sparse (see summary in Shibuya et al. 1992). It appears that eruption frequency and magnitudes have been sporadic during the last 140 ka. Smith & Allen (1993) have indicated that about 18 of the volcanoes, which include most of those associated with the largest eruption volumes, were formed between about 20 ka and 10 ka whilst Rangitoto, the largest and most recently formed volcano, was active about 600 yr BP (Robertson 1986).

Phreatomagmatic activity has occurred at about 75% of the volcanoes (Smith & Allen 1993) and is dominant at 34 centres (Houghton et al. 1991); fire fountaining has also occurred at about 75% of the centres (Smith & Allen 1993) and has resulted in some 30 scoria cones. Maars range up to about 1000 m in diameter whilst cones range up to about 700 m in basal diameter and 100 m in height. Rangitoto is anomalously large, having produced a volume of material approximately equal to the total volume of all previous eruptions.

The structural control on the location of the field as a whole or on individual centres is unknown. There is no clear pattern in the spacio-temporal distribution of the centres and therefore little to indicate the probability of where (or when) the next eruption is likely to occur. Magmas producing these volcanoes are thought to rise rapidly from mantle depths (Smith 1992) perhaps initiated by changes in the regional tensional stress field (cf. the Eifel field – Schmincke et al. 1983).

The city of Auckland, which contains about one third of the population of New Zealand, largely lies within the Auckland volcanic field. There is little historical record of eruptions in Auckland and hence the nature of activity within the field can be deduced only from geological and geophysical evidence. The greatest hazard that a future eruption in Auckland presents is that of a surge (Cassidy et al. 1986). Mapping of tephra deposits has shown that surges typically have travelled distances of up to 1.5 km, though tephra from Rangitoto volcano extends a distance of 12 km (Smith & Allen 1993). Other destructive but more restricted hazards in Auckland are presented by fire fountaining, air fall tephra, and lava flows which have reached up to 9.5 km in length.

In assessing the hazard that renewed volcanic activity may present, key factors are the size of any phreatomagmatic eruption and the possibility that more than one centre may be active at any one time. Rapid erosion of surge deposits and more recent urban development often make it difficult to assess the magnitude of previous phreatomagmatic activity at any given centre. Such activity will, of course, depend on the volume and recharge rate of the water body with which magma comes into contact, and the volume and rate of ascent of magma. Evidence for contemporaneous activity at more than one centre has not been found to date in the geological record (Rout et al. 1993).

Geophysical data

Figure 2 presents the results of new aeromagnetic data collected along 300 line-kilometres at an average elevation of 300 m over the southern part of the Auckland volcanic field. Flight lines are orientated NE-SW with an average spacing of 500 m; the horizontal sampling interval along the lines is 100 m. Positional control was determined by differential GPS, accurate to about ±5 m. Diurnal variations of the geomagnetic field were recorded at a local base station and subtracted from the survey data together with the International Geomagnetic Reference Field (IGRF) to give the total field anomaly data. Also shown in Fig. 2 (inset) is the aeromagnetic data over six volcanic centres previously surveyed by Rout et al. (1993). The combined dataset therefore represents two-fifths of the total number of centres and virtually all of those erupted within the youngest Plio-Pleistocene deposits.

The data clearly show numerous high-amplitude short-wavelength anomalies superimposed upon a more broad-wavelength regional field. All except one of these anomalies, correlate with volcanic centres although some centres are not associated with clearly defined anomalies. Both cones and maars are associated with strong anomalies (up to a maximum of 440 nT peak-to-peak; Table 1) indicating the presence of significant amounts of strongly magnetized material. Most anomalies are of positive polarity as would be expected given that the rocks were formed during the current normal geomagnetic polarity Brunhes epoch (0 to 0.7 Ma). Suprisingly, however, three centres have negative anomalies (Puketutu, Crater Hill and Wiri). It should be noted that remanent magnetization is significantly greater than induced in these basalts (Rout 1991) and therefore total magnetizations (and hence magnetic anomalies) reflect those acquired during the formation of the volcanoes.

Structural styles

Pukaki Crater and Crater Hill, two maars separated by only 1.5 km, have tuff rings of similar dimensions (Fig. 1), with outer diameters of about 1.5 km and heights of 10–20 m (Rout 1991). At Pukaki Crater the tuff is dominated by comminuted sedimentary material, but the presence of liquid magma is indicated by the occurrence of lithic clasts coated with dense basalt rinds. The material at the base of the tuff ring is poorly bedded and has a 'plastered' appearance whereas at higher levels, bedding (including cross-bedding) is clearer, suggesting that water was present throughout the eruption but that the volume available decreased during the event.

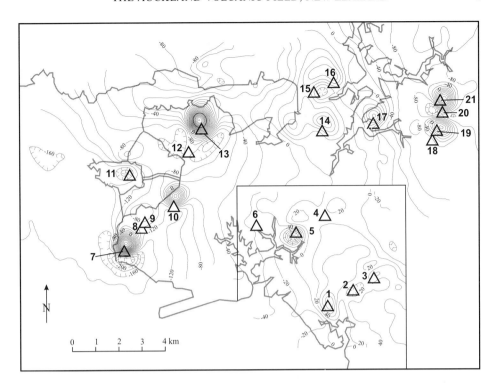

Fig. 2. Total force aeromagnetic map of the southern part of the Auckland volcanic field (as defined in Fig. 1) at a flight elevation of 300 m. Contour interval is 20 nT, tick marks on contours denote negative anomaly centres. Inset shows data from Rout *et al.* (1993), flown at a height of 240 m. Triangles denote volcanic centres with index numbers as in Fig. 1.

Crater Hill shows a more complex eruption sequence consisting of four near-concentric landforms (Houghton *et al.* 1986, 1991): an outer tuff ring, an inner scoria rampart, a low lava shield and a small scoria cone. The outer tuff ring is constructed from an early phreatomagmatic phase and a later magmatic stage which are faulted and overlain by a younger phreatomagmatic unit; three magmatic units make up the inner scoria rampart (Houghton *et al.* 1986, 1991). They consider that maar collapse may have terminated the first magmatic phase, renewed phreatomagmatic activity then cleared the vent before further magmatic activity occurred. Thus, in this case it appears that the water supply was not constant throughout the eruption. The lava shield is interpreted as a remnant of magma filling and subsequent draining from the maar (Houghton *et al.*, 1986).

Geophysical data provide further evidence for the contrasting eruptive histories at Pukaki Crater and Crater Hill despite their proximity and similar dimensions. Pukaki Crater is characterized by a small negative residual gravity anomaly (-0.6 mGal) and no associated magnetic anomaly whereas Crater Hill is associated with a positive residual gravity anomaly ($+2.6$ mGal) and a significant magnetic anomaly (240 nT peak-to-peak amplitude). These anomalies have been interpreted (Rout *et al.* 1993) in terms of a shallow (25 m) dish-shaped depression below Pukaki Crater filled with low density mud and colluvium and, in the case of Crater Hill, an extensive solid basalt body (volume 0.02 km^3) extending to 120 m below ground level. This basalt body underlying the Crater Hill centre is interpreted as basalt

Table 1. Physical dimensions and magnetic parameters of volcanoes in the Auckland field

Volcano	Maar diameter (m)	Cone diameter (m)/ height (m)	Magnetic anomaly amplitude peak-to-peak (nT)	Magnetic anomaly polarity/magnetization direction[2]
1. McLaughlins Hill		300/40	140[1]	positive/normal
2. Wiri		300/50	40[1]	negative/N down
3. Ash Hill	200		60[1]	positive/-
4. Kohuora	400		40[1]	positive/-
5. Crater Hill	600	50/10	220[1]	negative/N down
6. Pukaki Crater	600		0[1]	na/-
7. Maungataketake		500/60	440	positive/normal
8. Otuataua		250/40	30	positive/normal
9. Pukeiti		100/20		positive/-
10. Waitomokia	700	50/20	260	positive/normal
11. Puketutu		600/60	130	negative/N down
12. Mangere Lagoon	600	50/10	0	na/-
13. Mt Mangere		600/70	400	positive/normal
14. Robertson Hill	600		80	positive/normal
15. Mt Richmond	400	200/40	110	equi-polarity/-
16. McLennan Hills		400/20		equi-polarity/S up
17. Pukekiwiriki	500		90	positive/normal
18. Hampton Park		100/20	160	positive/W up
19. Otara Hill		300/50		positive/W up
20. Green Hill		300/50	260	positive/normal
21. Styaks Swamp	200			positive/normal

Table 1 shows the relationship between physical dimensions and magnetic parameters for volcanic centres in the southern half of the Auckland field. Volcano index numbers as in Fig. 1. Physical dimensions are approximate, based on Searle (1981). Some centres are characterized by both significant maar and cone structures; maars are identified by extensive circular depressions surrounded by elevated tuff rings. Cone diameters are defined at the level of the surrounding topographic surface (close to sea level for all centres). Magnetic anomaly is total force anomaly from Fig. 2 and magnetic polarity is the dominant polarity. na = not applicable; dash = unknown.
Normal direction means present-day geomagnetic inclination (i.e. about 60° upwards or negative), N down means magnetization vector is downwards (i.e. positive inclination) towards the north etc.
Bracketing of four pairs of adjacent centres indicates that separate anomalies are not resolved.
[1] Rout et al. (1993); [2] Shibuya et al. (1992).

infilling a maar-type crater that had been excavated to a depth of about 80 m, below which a narrowing of the modelled body may represent the upper portions of the feeder conduit.

Clearly the depth of excavation resulting from the phreatomagmatic activity at these two volcanoes is not related to the thickness of the Plio-Pleistocene sediments within which they have been erupted since the two volcanoes are adjacent and, indeed, in the case of Crater Hill the excavation must extend into the underlying Miocene sediments. Water in the young shallow sediments was probably involved initially at both volcanoes, and in the case of Pukaki Crater either the magma supply ceased or no deeper aquifer was encountered. In contrast, at Crater Hill a deeper aquifer in the Miocene sediments was encountered, possibly during a later phase of the eruption, forming the later phreatomagmatic deposits.

The aeromagnetic data (Fig. 2) show that other maars in the south of the field, namely Pukekiwiriki, Robertson Hill and Waitomokia, are associated with significant magnetic anomalies. This suggests that these volcanoes also have substantial basaltic masses concealed below them, similar to that below Crater Hill. Thus, deep excavation is not an uncommon scenario for maar development in this part of the field and may reflect the widespread occurrence of significant aquifers within the Miocene sediments. Similar deep excavations can be inferred from basalt bodies modelled beneath maars occurring within Miocene sediments in the northern half of the field (Nunns 1975) which further confirms that the process of deep (100 m) excavation by phreatomagmatic eruption is independent of the occurrence of Plio-Pleistocene surficial deposits at the site of eruption.

Eruption timing

Eruptions within monogenetic fields are generally considered to be restricted to only one volcanic centre at any one particular time (e.g. Cas & Wright 1987) and this has, until recently, been regarded as the case for Auckland. However, the occurrence of identical anomalous magnetization directions in three of the centres (Shibuya et al. 1992) suggests that they may have been active contemporaneously. The value of aeromagnetic data in this context is that they confirm an anomalous bulk magnetization for all three centres and hence greatly enhance the statistical significance of the limited palaeomagnetic measurements, especially for Puketutu where only a single site was sampled.

Given the similar ages determined for these three centres – Wiri: 25–28 ka; Crater Hill: 29 ka; Puketutu: 22 ka (Polarch et al. 1969; Grant-Taylor & Rafter 1971; Wood 1991) – the anomalous measured magnetization directions were attributed to an excursion of the geomagnetic field somewhere between 25 ka and 50 ka (Shibuya et al. 1992). Since the observed anomalous magnetization directions agree to within about $5°$ and excursion paths are thought to be rapid, at least $1° a^{-1}$ (Coe & Prevot 1989), i.e. at least an order of magnitude faster than current rates of secular variation, this provides very strong evidence for essentially concurrent eruption from these three centres. These conclusions are reinforced by geochemical data which provide evidence for a common source from a single magma batch (Rout et al. 1993).

Furthermore, Shibuya et al. (1992) report anomalous, though somewhat different, magnetization directions in three other centres (Otara Hill, Hampton Park and

McLennan Hills) which they attribute to the same excursion event. If this is the case and assuming that excusions are short-lived phenomenon, i.e. perhaps much less than a few hundred years (Mankinen *et al.* 1985; Levi & Karlin 1989), then the conclusion of broadly contemporaneous eruption at all six centres is a distinct possibility. Since Otara Hill and Hampton Park are closely adjacent then these may be vents of the same volcano. Currently, detailed magnetic modelling and geochemical studies are underway to evaluate the significance of anomalous magnetizations at these three centres.

The fact that the volcanic rocks have clearly recorded a geomagnetic excursion has a bearing on assessing the total duration of eruption at any one centre. This is normally a very difficult parameter to determine given the limited resolution of dating methods and is usually estimated from modern analogues to be relatively short, perhaps of the order of several years only (e.g. Cas & Wright 1987). In the case of Puketutu, Crater Hill and Wiri, because the measured and modelled bulk magnetizations are closely similar and can be attributed to an excursion then this provides evidence that these volcanoes were entirely built over a small number of years.

Further, though less strong, support for short eruption durations is provided by the apparent orientation of the typically bipolar anomalies (a parameter dependent on their bulk magnetic declinations) associated with a number of the volcanoes surveyed. If these volcanoes had been active over several hundred years for example, then given the normal rate of secular variation, bulk declinations (and hence the orientation of the resulting bipolar anomalies) would be expected to average at about zero. This is clearly not the case for a number of centres which exhibit significant easterly or westerly deflections of declination (e.g. McLaughlins Hill, Mt Mangere and Green Hill/Styaks Swamp – see Fig. 2) and indeed detailed magnetic modelling for McLaughlins Hill (Rout *et al.* 1993) gives a best-fit magnetic declination of about 20° west. To exhibit significantly non-zero bulk declinations implies a total duration of eruption which spans only a short time segment of the normal secular variation path, i.e. no more than about 100 years. This interpretation is also consistent with the assumed short lifetimes of Auckland's monogenetic volcanoes, though it may be noted that an eruption duration of several hundred years has been inferred for the anomalously large Rangitoto centre from palaeomagnetic studies of declination variation (Robertson 1986).

Conclusions

Aeromagnetic data are a particularly effective means of determining the substructure of small volcanic centres. Aeromagnetic data from the southern half of the Auckland volcanic field have shown that significant concealed basalt bodies must occur beneath many of the centres characterized by phreatomagmatic eruption; these are interpreted as basalt-filled maars.

There is no correlation between the surface geology, in particular the thickness of typically saturated young surficial sediments, and the occurrence of deep maars. Deep excavation may therefore be controlled by the occurrence of significant and, according to these data, widespread aquifers in the Miocene sediments. Given the common occurrence of substantial volumes of basalt beneath these centres, calculations of the total volume of magma generated by the Auckland volcanic field may be somewhat underestimated.

The similar anomalous bulk magnetization directions for the Crater Hill, Wiri and Puketutu centres provides evidence that these three volcanoes were active contemporaneously. In addition, there is some evidence that a further three centres (Hampton Park, Otara Hill and McLennan Hills) were active at or about the same time. The apparently erratic frequency of eruptions in the Auckland field is therefore emphasized here by the evidence for contemporaneous eruption; clearly the Auckland volcanic field is characterized by spasms of activity. Rangitoto, the most recent centre, may represent the spatially focused activity of such a spasm since it is anomalous in both size and duration and is the only eruption that has occurred in the last $c.7000$ years.

The Wiri, Crater Hill and Puketutu centres are aligned along an approximately NW–SE trend which is paralleled by a similar trend between the Otara Hill/Hampton Park and McLennan Hills centres. This orientation is close to that of the dominant regional NNW–SSE Late Tertiary–Quaternary structural trend (which is most apparent in basement rocks immediately to the east of the area), suggesting that the sites of contemporaneous activity may be structurally controlled by following pre-existing zones of weakness or even initiated by recent changes in SW–NE tensional stresses in the Auckland region. The geophysical data provide no evidence for subsurface connections between distinct centres, suggesting that either magma drained at the end of activity or any connection is too narrow to be resolved by the geophysical data.

The new result that possibly as many as six volcanic centres were active at one time, or at least within the same short period (perhaps lasting only a hundred years), is of considerable significance in terms of the hazard presented by the Auckland volcanic field and its likely impact on the local population. It also highlights the limitations of using simple calculations of the average return time for eruptions (which is at least 2000 years for Auckland's volcanoes) in hazard assessment. Although any one centre could be expected to affect only a limited area of the city, the combined effect of multiple active centres would be much greater and should be considered in planning responses to renewed activity. Another important factor in hazard assessment for Auckland is the frequency of occurrence of phreatomagmatic events that create maars of significant depth; these are often not apparent from surface mapping but are clearly evidenced by the geophysical data.

We are grateful to Auckland University Research Committee for financial support, Colin Yong for technical assistance and Louise Cotterall for assistance with manuscript preparation. CAM acknowledges BHP (Minerals) Ltd for financial support.

References

ALLEN, S. R. 1992. *Volcanic Hazard from the Auckland Volcanic Field*. MSc thesis, University of Auckland.

CAS, R. A. F. & WRIGHT, J. V. 1987. *Volcanic Successions, Modern and Ancient*. Unwin Hyman, London.

CASSIDY, J., LOCKE, C. A. & SMITH, I. E. M. 1986. Volcanic hazard in the Auckland region. *New Zealand Geological Survey Record*, **10**, 60–64.

COE, R. S. & PREVOT, M. 1989. Evidence suggesting extremely rapid field variation during a geomagnetic reversal. *Earth and Planetary Science Letters*, **92**, 292–298.

GRANT-TAYLOR, T. C. & RAFTER, T. A. 1971. New Zealand radiocarbon age measurements. *New Zealand Journal of Geology and Geophysics*, **14**(2), 364–402.

GUTMANN, J. T. 1979. Structure and eruptive cycle of cinder cones in the Pinacate volcanic field and the controls of Strombolian activity. *Journal of Geology*, **87**, 448–454.

HEMING, R. F. 1980. Patterns of Quaternary basaltic volcanism in northern North Island, New Zealand. *New Zealand Journal of Geology and Geophysics*, **23**, 335–344.

—— & BARNET, P. R. 1986. The petrology and petrochemistry of the Auckland volcanic field. *In*: Late Cenozoic volcanism in New Zealand. *Royal Society of New Zealand Bulletin*, **23**, 64–75.

HOUGHTON, B. F., WILSON, C. J. N., SMITH, I. E. M. & PARKER, R. J. 1986. Crater Hill. *International Volcanological Congress, February 1986, Handbook*. International Association of Volcanology and Chemistry of the Earth's Interior, Wellington, 69–83.

——, ——, —— & —— 1991. A Mixed Deposit of Simultaneously Erupting Fissure Vents: Crater Hill, Auckland, New Zealand. *New Zealand Geological Survey Record*, **43**, 45–50.

KERMODE, L. O. 1992. *Geology of the Auckland urban area. Scale 1:50000. Institute of Geological and Nuclear Science geological map 2*. Institute of Geological and Nuclear Sciences Ltd., Lower Hutt, New Zealand.

LEVI, S. & KARLIN, R. 1989. A sixty thousand year paleomagnetic record from Gulf of California sediments: secular variation, late Quaternary excursions and geomagnetic implications. *Earth and Planetary Science Letters*, **92**, 219–233.

MANKINEN, E. A., PRÉVOT, M., GROMMÉ, C. S. & COE, R. S. 1985. The Steens Mountain (Oregon) geomagnetic polarity transition, 1. Directional history, duration of episodes, and rock magnetism. *Journal of Geophysical Research*, **90**, 10393–10416.

NUNNS, A. 1975. *A geophysical investigation of Auckland explosion craters*. BSc(Hons) thesis, University of Auckland.

POLARCH, H. A., CHAPPELL, B. W. & LOVERING, J. 1969. A.N.U. radiocarbon age date list III. *Radiocarbon*, **11**(2), 245–280.

ROBERTSON, D. J. 1986. A paleomagnetic study of Rangitoto Island, Auckland, New Zealand. *New Zealand Journal of Geology and Geophysics*, **29**, 405–411.

ROUT, D. J. 1991. *A geophysical and volcanological study of the Wiri and Papatoetoe volcanoes, Auckland*. MSc(Hons) thesis, University of Auckland.

——, CASSIDY, J., LOCKE, C. A. & SMITH, I. E. M. 1993. Geophysical evidence for temporal and structural relationships within the monogenetic basalt volcanoes of the Auckland volcanic field, northern New Zealand. *Journal of Volcanology and Geothermal Research*, **57**, 71–83.

SCHMINCKE, H.-U., LORENZ, V. & SECK, H. A. 1983. The Quaternary Eifel Volcanic Fields. *In*: K. FUCHS, K. VON GEHLEN, H. MÄLZER, H. MURAWSKI & A. SEMMEL (eds) *Plateau Uplift. The Rhenish Shield – A Case History*. Springer Verlag, Berlin, pp. 139–151.

SEARLE, E. J. 1981. *City of Volcanoes: A geology of Auckland*. Longman Paul, Auckland, NZ.

SHIBUYA, H., CASSIDY, J., SMITH, I. E. M. & ITAYA, T. 1992. A geomagnetic excursion in the Bruhnes epoch recorded in New Zealand basalts. *Earth and Planetary Science Letters*, **111**(1), 41–48.

SMITH, I. E. M. 1989. North Island *In*: R. W. JOHNSON (ed.) *Intraplate Volcanism in Australasia*. Cambridge University Press, pp. 157–162.

—— 1992 Chemical zoning in small volume Basaltic Volcanoes in the Auckland Volcanic Field, Northern New Zealand: Evidence for Sub-crustal Fractionation Processes. *Geological Society of Australia Abstracts*, **32**, 207.

—— & ALLEN, S. R. 1993. *Volcanic hazards at the Auckland Volcanic Field*. Volcanic hazards information series no 5. Ministry of Civil Defence, Wellington, New Zealand.

SPORLI, K. B. 1989. Tectonic framework of Northland, New Zealand. *In*: SPORLI, K. B. & KEAR, D. (eds) Geology of Northland. *Royal Society of New Zealand, Bulletin*, **26**, 3–14.

WATERHOUSE, B. C. 1967. Groundwater Supply Auckland and Northland. *In*: Industrial Minerals and Rocks 1966. New Zealand Department of Scientific and Industrial Research, Information Series, **63**, 24–28.

WOOD, I. A. 1991. Thermoluminescence dating gives new ages for some Auckland basalts. Geological Society of New Zealand Miscellaneous Publications, **59A**, 147.

Impact of tephra fall and environmental change: a 1000 year record from Matakana Island, Bay of Plenty, North Island, New Zealand

TERESA M. GILES,[1] REWI M. NEWNHAM,[1] DAVID J. LOWE[2] & ADAM J. MUNRO[2]

[1] *Department of Geographical Sciences, University of Plymouth, Drake Circus, Plymouth PL4 8AA, UK*
[2] *Department of Earth Sciences and Geochronology Research Unit, University of Waikato, Private Bag 3105, Hamilton, New Zealand*

Abstract: Palynological evidence was used to determine the development of vegetation communities on Matakana Island, North Island, New Zealand, over the last 1000 radiocarbon years. The pollen record indicates that changes occurred in the vegetation immediately following fallout deposition of the Kaharoa Tephra approximately 100 km from source at *c.* 665 years BP. Such changes may be a direct response to the impact of tephra fall, although the possibility of anthropogenic disturbance cannot be discounted. As a result of the eruption some taxa (*Leucopogon fasciculatus* and *Tupeia antarctica*) became at least temporarily extinct from the area. Two phases of anthropogenic influence on the environment are recorded in the pollen record: Polynesian, followed by European inhabitation of the island, giving a detailed history of human influence in the area for the millennium.

The North Island of New Zealand has endured an extremely violent volcanic history throughout the Quaternary. Numerous thick pyroclastic deposits, including widespread fallout tephra layers, blanket the central North Island and there is a well established tephrostratigraphic record for the late Quaternary, i.e. the last *c.* 65 000 years (Lowe 1988; Froggatt & Lowe 1990; Wilson 1993; Alloway *et al.* 1995; Donoghue & Neall 1996; see also Newnham *et al.* 1999). The bulk of these deposits were erupted from the rhyolitic volcanic centres of Okataina, Taupo, Maroa and Mayor Island (Tuhua) and the andesitic centres of Tongariro and Egmont (Fig. 1C). Research on the impacts that eruptives from these volcanic centres have had on New Zealand vegetation communities has largely been restricted to localized areas in the North Island with work focusing on proximal volcanic impacts and both short- and long-term effects (e.g. Clarkson *et al.* 1988; McGlone *et al.* 1988; Clarkson 1990; Lees & Neall 1993; Clarkson & Clarkson 1994; Horrocks & Ogden 1998). Some research has also focused briefly on distal volcanic impacts. Palynological studies have shown evidence for the occurrence of forest fires following distal tephra fallout through increases in charcoal fragments and bracken spores immediately above tephra layers preserved in peat bogs and lake deposits in the North Island (McGlone 1981; Newnham *et al.* 1989, 1995a; Wilmshurst & McGlone 1996; Newnham & Alloway in press).

Human activity and deforestation began late in New Zealand with the arrival of

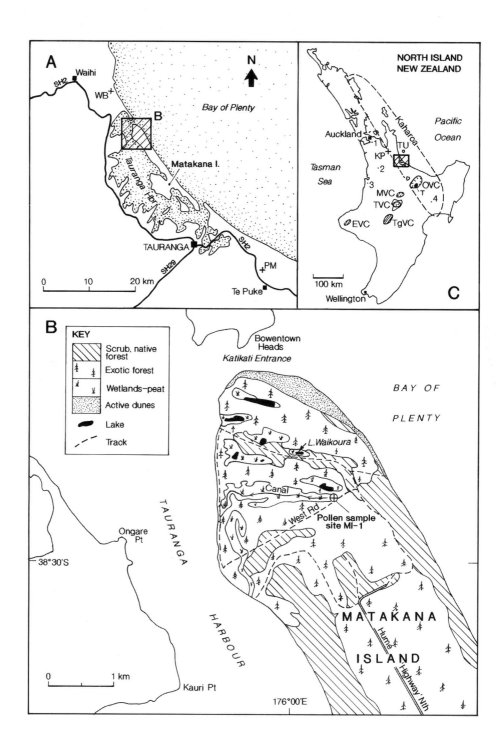

Polynesians around or after 1000 radiocarbon years BP, with the most severe anthropogenic impacts occurring between 730–500 years BP (McGlone 1989). (Note that throughout the text, ages are reported in terms of conventional radiocarbon years BP; calibrations of ages newly obtained in this study are listed in Table 2.) The combination of frequent and extensive volcanic activity, an excellent tephrostratigraphical record, and the short timespan of human activity provide great potential for studying the distal impacts of volcanism in New Zealand.

In this study, palynological changes in a peat core extracted from Matakana Island in the western Bay of Plenty region, North Island, New Zealand are examined. This work focuses particularly on vegetation changes following deposition of the Kaharoa Tephra *c*. 665 years BP. The preliminary results also reveal important insights into the environmental history of the island over the last 1000 years BP.

The study site

Matakana Island is a large barrier island, 26 km long, enclosing Tauranga Harbour (Fig. 1A). The main part of the island is composed primarily of Holocene sands (Marshall *et al*. 1994; Shepherd *et al*. 1997). The study site is situated at the northwestern end of the island adjacent to the Katikati Entrance of the harbour. This area comprises a dune-dominated landscape (all < 10 m above sea-level, mean high water spring) containing wetland areas and shallow dune lakes bordered by small peaty wetlands, all of which developed during the last *c*. 900 years (Munro 1994). Site MI-1 is located near the eastern end of a small (*c*. 0.5 km^2) low-lying interdune wetland adjacent to West Rd (Fig. 1B). A sluggish canal (*c*. 1 m wide near MI-1), constructed in the 1960s, partially controls the water table level of the wetland. The local catchment of this wetland is very limited, hence any inwashing is likely to be from adjacent dunes. The sampling site is currently dominated by grey willow (*Salix cinerea*) with toi-toi (*Cortaderia*), gorse (*Ulex*) and bracken (*Pteridium*) scrub, but is surrounded by exotic plantation forest (chiefly *Pinus radiata*) established initially in the mid-1920s. (Self-sown exotic *P. radiata* on the island dates from around 1900 to 1910.) Small areas of native forest and scrub still exist on the northwestern part of the island (Fig. 1B).

Archaeological evidence for human impact on Matakana Island has been found at a number of sites (Marshall *et al*. 1994). Pa (fortified village) sites, terraces and shell middens are less common on the northwestern end of the island in the vicinity of site MI-1, possibly because of burial or destruction by shifting sands and rapid erosion,

Fig. 1. (a) The western Bay of Plenty region showing the location of Matakana Island and study area adjacent to Tauranga Harbour. WB (Waihi Beach) and PM (Papamoa) refer to Holocene sites described by Newnham *et al*. (1995*b*). (b) The northwestern part of Matakana Island and the location of pollen sampling site MI-1. (c) The study area with respect to the rhyolitic Okataina Volcanic Centre (OVC) and Mt Tarawera volcano (T), the source of the 665 years BP Kaharoa Tephra. The dashed line is the 30 mm-thick isopach of Kaharoa Tephra (from Pullar *et al*. 1977). Other recently active volcanic centres shown are Tuhua (TU), Maroa (MVC), Taupo (TVC), Tongariro (TgVC), and Egmont (EVC) (Froggatt & Lowe 1990; Wilson *et al*. 1995). Numbers mark locations of pollen sites currently under investigation: 1, Kohuora crater (Auckland); 2, Lake Rotoroa (Hamilton); 3, Lake Taharoa; 4, Kaipo peat bog (Te Urewera National Park; see Lowe & Hogg 1986). KP marks the location of Kopouatai bog, as described by Newnham *et al*. (1995*a*).

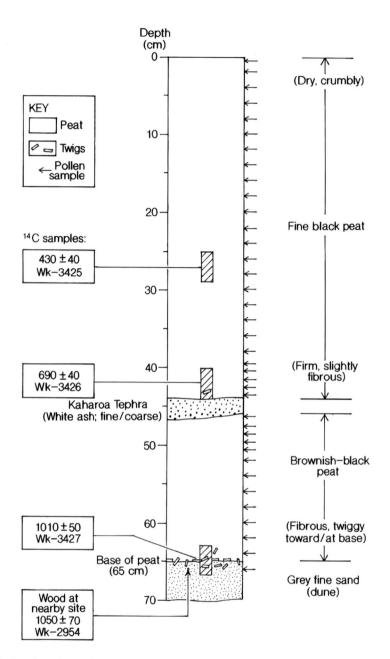

Fig. 2. Stratigraphy and chronology of peat profile for site MI-1, Matakana Island, at U13/758083 (grid references here and in Tables 1 & 2 are based on the metric 1:50 000 topographical map series NZMS 260), and pollen sampling positions. Ages are in conventional radiocarbon years BP (see Table 2).

Table 1. *Comparison of electron microprobe analyses of glass[a] in Kaharoa Tephra at Matakana Island with glass sampled at source (Mt Tarawera), Waihi Beach, and Kopouatai (Fig. 1)*

	Matakana Island[b]	Mt Tarawera[c]	Waihi Beach[d]	Kopouatai[e]
SiO_2	77.51 (0.19)	78.34 (0.21)	78.46 (0.40)	78.22 (0.31)
Al_2O_3	12.81 (0.15)	12.50 (0.12)	12.31 (0.11)	12.49 (0.15)
TiO_2	0.10 (0.03)	0.11 (0.02)	0.14 (0.06)	0.13 (0.04)
FeO^f	0.88 (0.10)	0.80 (0.08)	0.88 (0.14)	0.76 (0.26)
MgO	0.09 (0.04)	0.10 (0.01)	0.10 (0.05)	0.09 (0.13)
CaO	0.53 (0.07)	0.61 (0.05)	0.72 (0.17)	0.55 (0.06)
Na_2O	3.80 (0.09)	3.22 (0.15)	3.53 (0.16)	3.42 (0.19)
K_2O	4.12 (0.13)	4.21 (0.14)	3.78 (0.15)	4.22 (0.43)
Cl	0.18 (0.03)	0.16 (0.02)	0.15 (0.03)	0.14 (0.03)
Water[g]	2.31 (1.29)	3.05 (0.93)	2.18 (1.67)	0.93 (0.68)
n	11	11	9	10

[a] Means and standard deviations (in parentheses, calculated to one standard deviation) normalized to 100% loss-free. Analysed by Jeol JXA-733 Superprobe at the Analytical Facility, Victoria University of Wellington. All analyses used beam diameter 10–15 μm, current 8–nA, accel. voltage 15 kV (Froggatt 1983). Anal. 1 calculated from 11×2 s counts across the peak, curve integrated; anal. 2–4 calculated from 3×10 s counts at the peak, meaned. Analyses of TiO_2, MgO, and Cl were below detection in some shards; these values were omitted from the means. n refers to number of analyses (individual shards) in mean.
[b] Core at MI-1 (U13/758083; this study).
[c] Section at V16/177252 in Mt Tarawera crater (from Hodder *et al.* 1991).
[d] Core at U13/702174 in Waihi Beach swamp (from Newnham *et al.* 1995b).
[e] Core at T13/380197 in Kopouatai bog (from Hodder *et al.* 1991).
[f] Total Fe as FeO.
[g] Difference between original analytical total and 100.

or through logging operations. Marshall *et al.* (1994) and Shepherd *et al.* (1997) reported that there is no evidence of human settlement on Matakana Island prior to the deposition of Kaharoa Tephra.

Peat stratigraphy, chronology, sampling and pollen analysis

The stratigraphy at site MI-1 comprises 65 cm of peat overlying aeolian dune sands (Fig. 2). A macroscopic tephra-fall layer 2–3 cm thick occurs at 44–47 cm depth within the peat and is identified here as Kaharoa Tephra, a rhyolitic eruptive from Tarawera volcano in the Okataina Volcanic Centre situated *c.* 100 km southeast from Matakana Island (Fig. 1C). The layer, consisting of fine ash overlying coarse ash, is dominated by biotite (*c.* 85% of the magnetic fraction), which is a diagnostic mineral for this tephra (Froggatt & Lowe 1990). Kaharoa Tephra has been identified at other sites on Matakana Island (Munro 1994; Shepherd *et al.* 1997) and along the western Bay of Plenty coastline including peat bogs at Papamoa and Waihi Beach (Fig. 1A; Newnham *et al.* 1995b). The identification is confirmed through major element analysis of its constituent glass by electron microprobe (Table 1).

Froggatt & Lowe (1990) reported an error-weighted mean age of 770 ± 20 years BP (*n* = 15) for Kaharoa Tephra. However, Lowe & Hogg (1992) and Newnham *et al.* (1995a) suggested that it was erupted around, or possibly soon after, 700 years BP. The most recent age estimate, based on cluster analysis of 22 age determinations (Lowe *et al.* 1998), is an age of 665 ± 15 years BP (equivalent to *c.* 600 calendar years BP). This age is statistically identical to the age of 690 ± 40 years BP (Wk-3426)

Table 2. *Radiocarbon dates relevant to peat profile sampled and Kaharoa Tephra, Matakana Island*

Lab no.[a]	Sample material[b]	Depth[c] (cm)	$\delta^{13}C$ (‰)	Conventional age (years BP)[d]	Calibrated date AD[e] 1 standard deviation (sd)	Calibrated date AD[e] 2 standard deviations (sd)	Comments
Site MI-1 (U13/758083, see Fig. 2)							
3425	P	25–29	−28.3±0.2	430 ± 40	1445–1519(.73) 1594–1622(.27)	1438–1532(.58) 1547–1635(.42)	
3426	PW	40–44	−27.8±0.2	690 ± 40	1295–1326(.49) 1352–1362(.15) 1366–1389(.35)	1286–1334(.45) 1338–1403(.55)	Directly overlies KT[f]
3427	WP	63–67	−26.8±0.2	1010 ± 50	1005–1006(.01) 1019–1070(.44) 1081–1125(.40) 1136–1154(.15)	993–1163(.92) 1168–1193(.06) 1199–1207(.01)	Base of peat/sand interface
Site c. 10 m S of MI-1 (Edge of canal [U13/758082])							
2954	W	40–46	−28.2±0.2	1050 ± 70	905–966(.01) 982–1070(.61) 1082–1126(.27) 1136–1154(.11)	893–927(.07) 939–1209(.93)	Base of peat/sand interface[g]
Sawmill section (U14/847946)							
2820	C	70–71	−25.5±0.2	810 ± 140	1069–1083(.04) 1124–1137(.04) 1153–1327(.78) 1350–1390(.14)	1001–1011(.01) 1017–1426(.99)	Directly underlies KT

[a] University of Waikato Radiocarbon Dating Laboratory (prefix Wk-).
[b] P, Peat; PW, peat with twigs*; WP, twiggy* peat; W, large piece of wood (?root); C, fine charcoal fragments dominated by material from large podocarp trees and potentially with inbuilt age (R. Wallace, pers. comm. 1999). All samples were washed in hot 10% HCl, rinsed, and dried. Fine roots were removed from the peat samples. *Twigs are Podocarpaceae, probably young *Dacrycarpus dacrydioides* (P. J. de Lange, pers. comm. 1994).
[c] Below top of section.
[d] Old half-life basis (5568 yr)±1 sd; × by 1.03 to calculate on new half-life basis.
[e] Based on method B (probability distribution) of Stuiver & Becker (1993) after first subtracting the S. Hemisphere correction factor of 40 years (Vogel et al. 1993). Calibration program Rev. 3.04A (Mac test #6). Numbers in parentheses are relative contributions to probabilities.
[f] KT refers to Kaharoa Tephra.
[g] Equivalent stratigraphic position to Wk-3427.

obtained by us at MI-1 (Fig. 2; Table 2) and so we have adopted an age of 665 ± 15 years BP for the Kaharoa Tephra. Two further ^{14}C ages from MI-1 are consistent with the chronostratigraphic position of Kaharoa Tephra (Fig. 2). The basal twiggy peat is aged 1010 ± 50 years BP (Wk-3427), an age corroborated by the statistically-identical age of 1050 ± 70 years BP (Wk-2954) obtained on a piece of wood from an equivalent stratigraphic position at a site close by (Table 2).

The rates of peat accumulation are reasonably uniform and relatively fast, the mean for the entire profile being c. 0.61 mm a^{-1}. From c. 1000 years BP to 665 BP (base of peat to Kaharoa Tephra) the rate was c. 0.55 mm a^{-1}; from Kaharoa to 430 years BP it increased to c. 0.72 mm a^{-1}; and from 430 years BP to present day (assumed to be at surface), c. 0.63 mm a^{-1}. The last value is likely to be a minimum rate because of possible oxidation and shrinkage of peat at the surface in recent times through drainage.

Samples for pollen analysis were taken from contiguous slices 2 cm thick from the surface to a depth of 39 cm, and from 51 cm to 67 cm; in the interval from 39–51 cm depth the contiguous slices were 1 cm thick (i.e. above and below the Kaharoa Tephra layer; Fig. 2). This finer sampling resolution was adopted to examine any vegetational changes in the area resultant upon fall of the tephra material. From each contiguous slice of peat, 0.5 cm^3 samples were extracted for pollen preparations. Standard palynological procedures were employed (Moore *et al.* 1991), excluding HF acid treatment in order to preserve tephra-derived glass shards in samples. The addition of exotic *Lycopodium* spores permitted the calculation of pollen concentrations. Deteriorated pollen grains were also counted and included in the pollen diagram (Fig. 3). Classification of deteriorated grains was based on Cushing (1967) with crumpled, broken and degraded grains being recorded separately. High levels of deteriorated pollen grains are assumed to be associated with periods of environmental instability and catchment or regional (e.g. wind) erosion. Charcoal fragments and tephra-derived glass shards were counted and expressed in Fig. 3 as percentages of total dryland pollen (250 grains counted per sample) plus total number of glass shards or charcoal fragments counted per sample.

Preliminary results

Pollen data for the 1000 year time span of site MI-1 are presented in Fig. 3. Pollen zones were defined according to major changes in vegetation composition throughout the profile. The changes that are of major interest for this research are those in pollen zones MI1 and MI2 which respectively illustrate the vegetation composition before and just after the deposition of the Kaharoa Tephra.

MI1 (1000 to 665 years BP)

At the base of pollen zone MI1 indications of burning are evident with a peak in charcoal levels and the presence of *Pteridium* spores. This is short lived and does not appear to have had a major effect on the vegetation. *Phyllocladus* pollen rises to dominance along with increasing numbers of *Agathis*, *Dacrydium* and *Dacrycarpus* pollen. These last two species possibly invaded the wetland fringes and lower waterlogged sites (Newnham *et al.* 1989). A period of vegetation stability on the island can be inferred from the diagram, with northern conifer–angiosperm forest well established and *Metrosideros* trees and lianes prominent in the area. Pollen from small angiosperm trees and herbaceous taxa are also present in very small numbers, which may indicate local presence, but not on the site itself.

Towards the top of the zone, a change in dominant species occurs, with *Agathis* pollen increasing while *Phyllocladus* pollen begins to decline, along with *Dacrycarpus*. These changes may reflect competition between the species which has caused the decline of *Phyllocladus* with the younger trees receiving little light due to a close canopy of tall *Agathis* trees. *Agathis* and *Phyllocladus* trees favour drier sites and are tolerant of infertile soils (Newnham *et al.* 1989) whereas *Dacrydium* and *Dacrycarpus* trees thrive on moist, swampy ground and could have invaded the fringes of the wetland area. Deteriorated pollen grains, also present in small numbers, were possibly washed or blown into the catchment from the surrounding dunes during periods of increased run-off, and became incorporated into the peat. A relatively stable environment is inferred from the pollen assemblages represented in zone MI1.

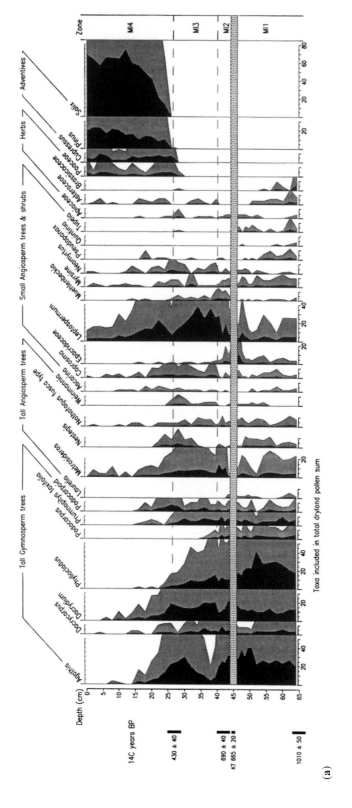

Fig. 3. Pollen diagram revealing changes in main taxa at site MI-1: (**a**) dryland pollen; (**b**) wetland pollen, spores and deteriorated grains. Silhouette graphs are presented with ×5 exaggeration. Total pollen concentration is expressed as total number of pollen grains and spores per cm³. Pollen represented as podocarpoid includes grains of *Prumnopitys*, *Podocarpus*, *Lagarostrobos* and *Lepidothamnus* pollen which were indistinguishable. KT represents Kaharoa Tephra. The pollen sampling interval is indicated by the vertical spacing between bars in the 'Total pollen & spore count per sample' curve.

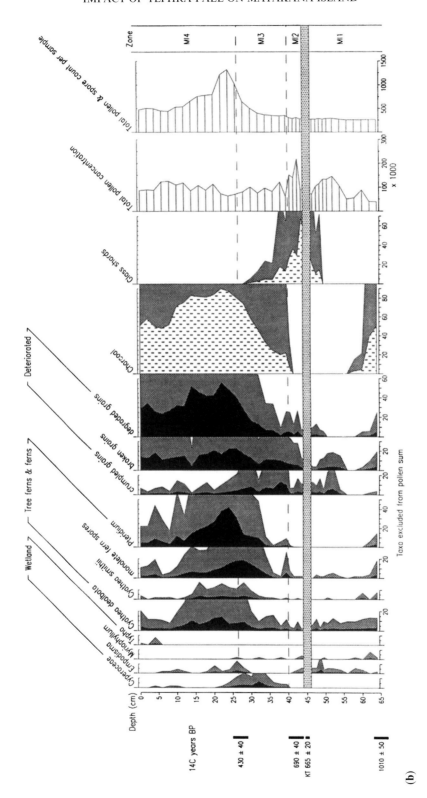

MI2 (Immediately post-Kaharoa Tephra, c. 665 years BP)

Immediately following deposition of the Kaharoa Tephra, numerous variations in the pollen spectrum point to significant disturbance in the catchment. Levels of broken and degraded pollen grains increase, along with a sharp rise in *Leptospermum* pollen from 7 to 17%. *Metrosideros* pollen levels also increase rapidly and, these species, along with *Leptospermum*, can be invaders following catchment disturbance. Some *Metrosideros* trees also thrive on fresh volcanic surfaces, along with *Dacrydium cupressinum* (McGlone *et al.* 1984; Newnham *et al.* 1989). *Muehlenbeckia* pollen levels rise above the Kaharoa Tephra layer, and may also indicate invasion by these shrubs or lianes onto the tephra surface.

Phyllocladus, Agathis and *Dacrydium* pollen levels decline gradually towards the top of the zone. It is hard to decipher whether the decline in these well-established trees is volcanically induced, as *Phyllocladus* was previously in decline before the Kaharoa eruption. *Agathis* and *Dacrydium* pollen levels follow a similar pattern after the deposition of the Kaharoa Tephra. Initially there is no reaction, but between 42 and 43 cm depth (1–2 cm above the tephra layer) there appears to be a significant decline which occurs a little later in *Agathis* trees. There then follows a very brief period of recovery into zone MI3, until eventually both species are eliminated during the European era (zone MI4). The initial short-lived decline could be attributed to a volcanic cause as it occurs soon after the tephra deposition and is not sustained. In addition, *Podocarpus* also decreases and eventually disappears after deposition of the tephra layer. However, this apparent decline may be an artefact of general deterioration in pollen quality because the taxon podocarpoid (which may include *Podocarpus* and other bisaccate pollen with non-radial sac pattern, and is used when generic distinction cannot be made) increases as *Podocarpus* pollen levels fall.

The hemiparasitic shrub *Tupeia antarctica*, which was growing in proximity to the site before the tephra was deposited, indicated by small amounts of pollen, apparently disappears after the eruption. However, the presence of just one pollen grain from this taxon at approximately 27 cm depth suggests that it may still have been present in the area, but perhaps confined to the mainland in more sheltered locations. *Tupeia* may have been striving to become established in the area, but the deposition of acid-laden tephra may have inflicted stress on the plant, contributing to its decline in the catchment. Similarly, Epacridaceae (mostly *Leucopogon fasciculatus*) pollen began a rapid increase at the top of zone MI1 but, following the Kaharoa Tephra deposition, this taxon declines to insignificant levels. This decline may also suggest possible plant damage by tephra accumulation inhibiting further expansion in the area.

MI3 (665 to 150 years BP)

Charcoal reappears at the top of zone MI2, first at low levels, but rapidly increasing in zone MI3. Minimum counts of *Agathis* pollen coincide with the appearance of charcoal. However, *Agathis* rises to former levels towards the top of the zone only to decline and eventually disappear in zone MI4. *Leptospermum* becomes dominant, with pollen values fluctuating around 40% and *Metrosideros* is also prominent but declines slightly following the peak in values immediately above the tephra. Cyperaceae, *Pteridium* and other ferns become established in the area as the charcoal trend rises. This trend, together with the high numbers of *Leptospermum* pollen and high charcoal levels, indicates invasion of bared surfaces following fire. Increases in damaged pollen grains provide further evidence for run-off and possibly wind erosion due to reduction in forest cover. This evidence, considered together with the inferred age of this zone, points to burning and deforestation for various purposes by early Polynesian settlers in the vicinity.

MI4 (150 years BP to present day)

This zone marks the beginning of the European era. All previously dominant native taxa decline and a significant change in vegetation composition occurs. Wetland plants, tree ferns and *Leptospermum* remain in lower numbers. The abundance of charcoal, together with degraded and broken pollen grains, points to continued catchment clearance by burning of the native vegetation and associated catchment erosion and run-off. Peaks in *Pteridium* spores and charcoal levels coincide, indicating that bracken probably colonized bare surfaces produced by fire. Rapid expansion of adventive species follows because *Salix* becomes dominant, with *Pinus* and *Cupressus* also present in significant numbers. Charcoal levels remain high to the top of the profile indicating sustained extensive clearance of the area up to recent times.

Discussion

Table 3 provides a summary of events at site MI-1 over the last 1000 radiocarbon years. These preliminary results indicate that significant changes occurred after the deposition of the Kaharoa Tephra. The interpretations are put forward with caution and thus cannot yet be attributed exclusively to tephra fallout because other factors, particularly anthropogenic, may have contributed to the changes displayed in the pollen diagrams. The rapid decreases in *Tupeia* and Epacridaceae seem to be connected with the deposition of tephra. It is possible that these shrubs were subjected to acid damage following tephra fall, and/or tephra accumulation on plant leaves, interfering with vital mechanical processes needed for plant survival. This initial damage may have weakened these plants, leaving them susceptible to disease, storm and wind damage and competitive pressure from other species. Increases in pollen of *Leptospermum*, *Metrosideros* and *Muehlenbeckia* may suggest subsequent invasion of canopy openings and bared surfaces by these shrubs and lianes following the decline of the previous vegetation occupying these sites (e.g. *Phyllocladus*, *Agathis*, Epacridaceae, *Tupeia*).

The distribution of tephra-derived glass shards immediately above and below the main Kaharoa Tephra layer (Fig. 3) indicates some disturbance in the profile. The spread of shards above the layer may point to increased run-off or wind erosion in the area as *Agathis* and *Dacrydium* trees began to decline, possibly creating gaps in the forest canopy and bared surfaces. Thus the shards may have been washed or blown in from the catchment. This conclusion is supported by increases in deteriorated pollen grains which also suggest catchment erosion and increased run-off or wind erosion. Glass shards are found almost to the top of zone MI3, and may be the result of burning and clearance by early Polynesian settlers causing continued erosion of the area.

Table 3. *Summary of main vegetation changes occurring on Matakana Island over 1000 ^{14}C year timespan.*

Years BP	Pollen zone	Dominant pollen taxa	Vegetation and environment
[a]150–0	MI4	*Salix, Pinus, Pteridium*	European era. Abundant charcoal and damaged pollen grains indicate extensive burning of native species and subsequent catchment erosion.
665–150	MI3	*Leptospermum, (Agathis)*	Recovery of *Agathis* and *Dacrydium*. Polynesian era: charcoal rise, *Pteridium* and Cyperaceae become established.
c. 665 (immediately post-Kaharoa Tephra)	MI2	*Leptospermum, Agathis*	Possible ash-fall induced disturbance. Local disappearance of *Tupeia, Quintinia*, Epacridaceae. Decline of *Agathis* and *Dacrydium*. Invasion by seral species.
1000–665	MI1	*Phyllocladus, Agathis*	Dominant northern conifer–angiosperm forest. Stable.

[a]Age of pollen zone MI4 based on appearance of adventive pollen rather than Wk-3425.

Glass shards have been disseminated below the tephra layer. Such dissemination may relate to the irregular surface of the peat bog on which the ash was deposited, with tephra being reworked physically to accumulate in hollows created by uneven vegetation cover. This would then enable the shards to be displaced below the main tephra deposit and become compressed into the peat following further sedimentation and compaction. Minor bioturbation and possible infiltration down root channels following tephra deposition may also have contributed to the downward spread of shards in the profile, as described for Kopouatai bog by Hodder *et al.* (1991). Further evidence for possible downwards dislocation of microscopic-sized particles is indicated at the base of pollen zone MI4, the European era, defined by the first appearance of adventive pollen (Fig. 3). This era spans the last *c.* 150 years of New Zealand history, yet the peat immediately beneath this zone is dated at 430 ± 40 years BP (Wk-3425). Either the date is in error, or adventive pollen grains have moved down the peat profile, as has been reported at other pollen sites (Newnham *et al.* 1995a).

The marked increase in total pollen concentration levels just above the Kaharoa Tephra (Fig. 3) could relate to two main factors. First, the delayed decline in pollen levels of *Agathis* and *Dacrydium* trees following the eruption could be due to survival mechanisms whereby the trees are in decline but are increasing pollen production levels in order to create new seedlings to replace dying trees. Eventually the trees may have become overwhelmed and weakened by tephra accumulation on leaves and branches with possible acid damage, hence causing temporary decline in both trees and pollen concentrations.

Alternatively, drying of the peat bog through the addition of the 2–3 cm-thick tephra layer may have resulted in slower peat accumulation for the period immediately following the eruption. Meanwhile, airborne pollen influx onto the peat surface may have remained similar to pre-eruption levels, thereby resulting overall in higher pollen levels per cubic centimetre of peat because of a slower rate of accumulation. A temporary slowing in peat accumulation, together with the new source of pollen derived from long distance dispersal by wind, and overland flow, as indicated by the increase in glass shards and deteriorated pollen counts, could explain the short-lived peak in pollen concentrations soon after the tephra layer was deposited.

Pollen zone MI3 is interpreted as the beginning of the Polynesian era with extensive burning in the area, indicated by the rapid rise in charcoal levels and gradual increases in *Pteridium* spores. It appears from Fig. 3 that the commencement of the Polynesian era begins a little later on Matakana Island than in other areas studied around the Bay of Plenty coast and eastern Waikato regions (Newnham *et al.* 1995a,b, 1998). At Papamoa and Waihi Beach (Fig. 1A), *Pteridium* and charcoal levels rise above the Kaharoa Tephra as tree pollen declines. However, adventive pollen was also found at this level, which Newnham *et al.* (1995b) suggested may have been due to contamination within the profile. Thus, no definite conclusions were drawn for either human or tephra impact following Kaharoa Tephra deposition at these sites. At Kopouatai bog (Fig. 1C), increases in *Pteridium* and charcoal commenced just below the Kaharoa Tephra and were interpreted as the result of regional human influence (Newnham *et al.* 1995a). It is possible that changes in the vegetation composition in zone MI2 at Matakana Island were the result of volcanic rather than human impact because charcoal and bracken levels do not appear directly above the tephra layer, but a further 4 cm above the boundary. However, the subsequent (presumably anthropogenic) disturbance may be superimposed on any long-term tephra fall impact on the vegetation, so that the extent of damage caused from the deposition of the Kaharoa Tephra cannot readily be seen from this diagram. It is also possible that initial human impacts on Matakana Island were not associated with burning. Thus post-Kaharoa vegetation changes could have been the result of a small population occupying the area exploiting local resources, thus producing a small, but palynologically discernible impact. Further work is required before firmer conclusions can be drawn.

Further work

The use of pollen analysis alone is insufficient to determine the precise effect tephra deposition has had on palaeoenvironments. Geochemical analysis will be undertaken to complement the results already obtained at Matakana Island, in particular to provide a clearer picture of catchment disturbance following deposition of the Kaharoa Tephra as inferred from the pollen data. Further palynological examination of vegetation changes, and geochemical analysis of sediments above earlier tephras deposited prior to human settlement in New Zealand, are required in order to isolate disturbances resulting from tephra-fall impacts.

This site is the first of a suite of tephra sequences from Auckland, Waikato and Bay of Plenty regions of New Zealand to be examined using fine-resolution palynology and geochemistry. These tephra-rich organic sequences provide opportunities to examine the extent to which the vegetation and environmental changes following tephra fall detected at Matakana Island occurred at other sites, and should provide detailed accounts of possible environmental disturbance which could be attributed to the distal impacts of tephra fall from volcanic eruptions in New Zealand.

Conclusions

Because the palynological signals for anthropogenic and volcanogenic disturbance may be similar, it is not yet possible to be certain as to whether vegetation changes occurring above the Kaharoa Tephra layer at Matakana Island are caused by distal volcanic impacts, human activity, or both. However, tentative conclusions inferred from the results show that the deposition of ash-grade fallout tephra may have caused some environmental disturbance on Matakana Island. This is supported by the rapid decline of certain taxa (*Leucopogon fasciculatus* and *Tupeia antarctica*) and the (somewhat delayed) decline in *Agathis* and *Dacrydium* pollen following tephra fall, as well as significant increases in pollen concentration, deteriorated pollen and microscopic tephra-derived glass shards. This evidence is compatible with that presented by Blackford *et al.* (1992) and Charman *et al.* (1995) who suggested that acid loading associated with fallout tephra deposition is a possible cause for palaeoenvironmental change in northern Britain.

The results presented in this report are preliminary, and further work will be undertaken to produce a more detailed record of events at site MI-1. Results of geochemical analysis and discussion of human settlement on Matakana Island will be presented at a later date.

This work forms part of a PhD project undertaken by TMG and was funded chiefly by the University of Plymouth, UK, DevR Fund, with support from the Department of Earth Sciences, University of Waikato, New Zealand. We acknowledge help from John Grattan with the identification of tephric glass shards and for comments on the manuscript.

We are grateful to the British Council (Wellington, New Zealand) for funding towards travel costs associated with this project through a Higher Education Link between Waikato and Plymouth universities. We also acknowledge funding for radiocarbon dating from the New Zealand Lottery Grants Board (DJL). Jim Dahm (Environment Waikato) and John Green (University of Waikato) helped with field work on Matakana Island, and Peter de Lange (Department of Conservation) identified wood fragments in the peat. Doug Sutton (University of Auckland) and Bruce McFadgen (Department of Conservation) kindly provided copies of the reports on archaeological and geomorphological work on the island, and Tom Higham (University of Waikato Radiocarbon Dating Laboratory) calculated the ^{14}C calibrations.

The paper is an output from the REPUTE ('Reconstructing palaeoenvironments using tephrochronology') Working Group of the Commission on Tephrochronology and Volcanology, International Union for Quaternary Research.

References

ALLOWAY, B. V., NEALL, V. E. & VUCETICH, C. G. 1995. Late Quaternary (post 28 000 year BP) tephrostratigraphy of north-east and central Taranaki, New Zealand. *Journal of the Royal Society of New Zealand*, **25**, 385–458.

BLACKFORD, J. J., EDWARDS, K. J., DUGMORE, A. J., COOK, G. T. & BUCKLAND, P. C. 1992. Icelandic volcanic ash and the mid-Holocene Scots pine (*Pinus sylvestris*) pollen decline in northern Scotland. *The Holocene*, **2**, 260–265.

CHARMAN, D. J., WEST, S., KELLY, A. & GRATTAN, J. P. 1995. Environmental response to tephra deposition in the Strath of Kildonan, northern Scotland. *Journal of Archaeological Science*, **22**, 799–809.

CLARKSON, B. D. 1990. A review of vegetation development following recent (<450 years) volcanic disturbance in North Island, New Zealand. *New Zealand Journal of Ecology*, **14**, 59–71.

—— & CLARKSON, B. R. 1994. Vegetation decline following recent eruptions on White

Island (Whakaari), Bay of Plenty, New Zealand. *New Zealand Journal of Botany*, **32**, 21–36.

CLARKSON, B. R., PATEL, R. N. & CLARKSON, B. D. 1988. Composition and structure of forest overwhelmed at Pureora, central North Island, New Zealand, during the Taupo eruption (*c.* AD 130). *Journal of the Royal Society of New Zealand*, **18**, 417–436.

CUSHING, E. J. 1967. Evidence for differential pollen preservation in late Quaternary sediments in Minnesota. *Review of Palaeobotany and Palynology*, **4**, 87–101.

DONOGHUE, S. L. & NEALL, V. E. 1996. Tephrostratigraphic studies at Tongariro Volcanic centre, New Zealand: an overview. *Quaternary International*, **34–36**, 13–20.

FROGGATT, P. C. 1983. Toward a comprehensive Upper Quaternary tephra and ignimbrite stratigraphy in New Zealand using electron microprobe analysis of glass shards. *Quaternary Research*, **19**, 188–200.

—— & LOWE, D. J. 1990. A review of late Quaternary silicic and some other tephra formations from New Zealand: their stratigraphy, nomenclature, distribution, volume, and age. *New Zealand Journal of Geology and Geophysics*, **33**, 89–109.

HODDER, A. P. W., DE LANGE, P. J. & LOWE, D. J. 1991. Dissolution and depletion of ferromagnesian minerals from Holocene tephra layers in an acid bog, New Zealand, and implications for tephra correlation. *Journal of Quaternary Science*, **6**, 195–208.

HORROCKS, M. & OGDEN, J. 1998. The effects of the Taupo Tephra eruption of *c.* 1718 BP on the vegetation of Mt Hauhangatahi, Central North Island, New Zealand. *Journal of Biogeography*, **25**, 649–660.

LEES, C. M. & NEALL, V. E. 1993. Vegetation response to volcanic eruptions on Egmont volcano, New Zealand, during the last 1500 years. *Journal of the Royal Society of New Zealand*, **23**, 91–127.

LOWE, D. J. 1988. Stratigraphy, age, composition, and correlation of late Quaternary tephras interbedded with organic sediments in Waikato lakes, North Island, New Zealand. *New Zealand Journal of Geology and Geophysics*, **31**, 125–165.

—— & HOGG, A. G. 1986. Tephrostratigraphy and chronology of the Kaipo Lagoon, an 11 500 year-old montane peat bog in Urewera National Park, New Zealand. *Journal of the Royal Society of New Zealand*, **16**, 25–41.

—— & —— 1992. Application of new technology liquid scintillation spectrometry to radiocarbon dating of tephra deposits, New Zealand. *Quaternary International*, **13/14**, 135–142.

——, MCFADGEN, B. G., HIGHAM, T. F. G., HOGG, A. G., FROGGATT, P. C. & NAIRN, I. A. 1998. Radiocarbon age of the Kaharoa Tephra, a key marker for late Holocene stratigraphy and archaeology in New Zealand. *The Holocene*, **8**, 499–507.

MARSHALL, Y., PAAMA, M., SAMUELS, M., SUTTON, D. & TAIKATO, T. 1994. *Archaeological survey of Matakana Island forest compartment 3, stands 2–4; compartment 12, stands 1–5; compartment 14, stand 2 and compartment 20, stands 1–5. Final report.* Auckland Uniservices Ltd, University of Auckland, Auckland.

MCGLONE, M. S. 1981. Forest fire following Holocene tephra fall. *In:* HOWARTH, R., FROGGATT, P. C., VUCETICH, C. G. & COLLEN, J. D. (eds) *Proceedings of Tephra Workshop*, Geology Department, Victoria University of Wellington Publication, **20**, 80–86.

—— 1989. The Polynesian settlement of New Zealand in relation to environmental and biotic changes. *New Zealand Journal of Ecology Supplement*, **12**, 115–129.

——, HOWARTH, R. & PULLAR, W. A. 1984. Late Pleistocene stratigraphy, vegetation and climate of the Bay of Plenty and Gisborne regions, New Zealand. *New Zealand Journal of Geology and Geophysics*, **27**, 327–350.

——, NIALL, V. E. & CLARKSON, B. D. 1988. The effect of recent volcanic events and climate change on the vegetation of Mt Egmont (Mt Taranaki), New Zealand. *New Zealand Journal of Botany*, **26**, 123–144.

MOORE, P. D., WEBB, J. A. & COLLINSON, M. E. 1991. *Pollen Analysis, (2nd edn)*, Blackwell Scientific Publications, London.

MUNRO, A. J. 1994. *Holocene evolution and palaeolimnology of a barrier spit, northwestern Matakana Island, Bay of Plenty, New Zealand.* MSc thesis, Library, University of Waikato, Hamilton, New Zealand.

NEWNHAM, R. M. ALLOWAY, B. V. A. in press. The Last Interglacial/Glacial cycle in Taranaki, western North Island, New Zealand: regional synthesis and new sites. *Palynology*.
——, DE LANGE, P. J. & LOWE, D. J. 1995a. Holocene vegetation, climate and history of a raised bog complex, northern New Zealand, based on palynology, plant macrofossils and tephrochronology. *The Holocene*, **5**, 81–96.
——, LOWE, D. J. & ALLOWAY, B. V. 1999. Volcanic hazards in Auckland, New Zealand: a preliminary assessment of the threat posed by central North Island silicic volcanism based on the Quaternary tephrostratigraphical record. *This volume*.
——, —— & GREEN, J. D. 1989. Palynology, vegetation and climate of the Waikato lowlands, North Island, New Zealand, since 18 000 years ago. *Journal of the Royal Society of New Zealand*, **19**, 127–150.
——, ——, MCGLONE, M. S., WILMSHURST, J. M. & HIGHAM, T. F. G. 1998. The Kaharoa Tephra as a critical datum for earliest human impact in northern New Zealand. *Journal of Archaeological Science*, **25**, 533–544.
——, —— & WIGLEY, G. N. A. 1995b. Late Holocene palynology and palaeovegetation of tephra-bearing mires at Papamoa and Waihi Beach, western Bay of Plenty, North Island, New Zealand. *Journal of the Royal Society of New Zealand*, **25**, 283–300.
PULLAR, W. A., KOHN, B. P. & COX, J. E. 1977. fallout Kaharoa Ash and Taupo Pumice, and sea rafted Loisels Pumice, Taupo Pumice, and Leigh Pumice in northern and eastern parts of the North Island, New Zealand. *New Zealand Journal of Geology and Geophysics*, **20**, 697–717.
SHEPHERD, M. J., MCFADGEN, B. G., BETTS, H. D. & SUTTON, D. G. 1997. *Formation, landforms and palaeoenvironments of Matakana Island, Bay of Plenty, and implications for archaeology*. Science and Research Series, **102**.
STUIVER, M. & BECKER, B. 1993. High-precision decadal calibration of the radiocarbon timescale, AD 1950–6000 BC. *Radiocarbon*, **35**, 35–65.
VOGEL, J. C., FULS, A., VISSER, E. & BECKER, B. 1993. Pretoria calibration curve for short-lived samples, 1930–3350 BC. *Radiocarbon*, **35**, 73–85.
WILMSHURST, J. M. & MCGLONE, M. S. 1996. Forest disturbance in the central North Island, New Zealand, following the 1850 BP Taupo eruption. *The Holocene*, **6**, 399–411.
WILSON, C. J. N. 1993. Stratigraphy, chronology, styles and dynamics of late Quaternary eruptions from Taupo Volcano. *Philosophical Transactions of the Royal Society London*, **A343**, 205–306.
——, HOUGHTON, B. F., MCWILLIAMS, M. O., LANPHERE, M. A., WEAVER, S. D. & BRIGGS, R. M. 1995. Volcanic and structural evolution of Taupo Volcanic Zone, New Zealand: a review. *Journal of Volcanology and Geothermal Research*, **68**, 1–28.

Volcanic hazards in Auckland, New Zealand: a preliminary assessment of the threat posed by central North Island silicic volcanism based on the Quaternary tephrostratigraphical record

REWI M. NEWNHAM[1], DAVID J. LOWE[2] & BRENT V. ALLOWAY[3]

[1] *Department of Geographical Sciences, University of Plymouth, Plymouth PL4 8AA, UK*
[2] *Department of Earth Sciences, University of Waikato, Private Bag 3105, Hamilton, New Zealand*
[3] *Department of Geology, University of Auckland, Tamaki Campus, Private Bag 92-019, Auckland, New Zealand*
Present address: Institute of Geological and Nuclear Sciences Ltd, Gracefield Research Centre, 69 Gracefield Road, PO Box 30–368, Lower Hutt, New Zealand

Abstract: The City of Auckland (population *c.* 1 million), built on a basaltic volcanic field active as recently as *c.* AD 1400, faces an additional volcanic threat: that from several large and productive rhyolitic and andesitic eruptive centres of the central North Island, 140–280 km to the south and southeast. Non-basaltic tephra fallout layers originating from these distal eruptive centres are numerous and widespread in the Auckland region and have primary thicknesses ranging from *c.* 1 mm to $\geqslant = 0.6$ m; ignimbrites up to 9 m thick are also documented but are uncommon. The assessment of volcanic hazards in Auckland is made problematical by the different types of volcanic threat posed by these two spatially distinct source areas, and by the lack of recognition hitherto given to the threat from the distal sources. This paper reviews the Quaternary records of distal volcanism affecting Auckland and outlines current investigations into the assessment of environmental impacts of past eruptions. Our preliminary results indicate that the potential threat to Auckland from the distal volcanic sources has been underestimated and that further research into the impacts of Quaternary volcanism on Auckland's environment and infrastructure is essential. The importance of this threat was underscored in mid-1996 when a small magnitude eruption of Mt Ruapehu necessitated closure of Auckland International Airport for three nights.

Assessments of volcanic hazards for the City of Auckland have focused entirely on the threat from the local volcanic field (Figs 1, 2). The surface geology of the city is dominated by Late Quaternary basaltic volcanism, represented by 49 centres, with the most recent eruption (Rangitoto) about 500 radiocarbon years ago (Brothers & Golson 1959; Nichol 1992; Lowe *et al.* 1999*b*) attesting to continuing activity. (All ages reported in the text are conventional radiocarbon ages based on the old half-life unless otherwise stated.) The field is characterized by low-frequency, low-magnitude

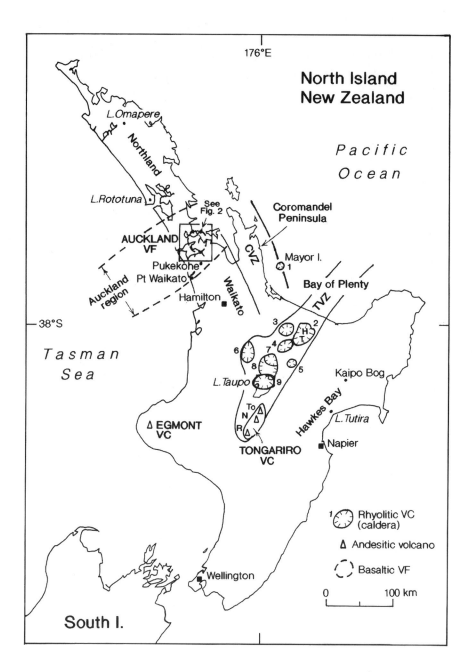

Fig. 1. Locations of the main volcanic centres active in the central North Island during the Quaternary (after Houghton et al. 1995; Wilson et al. 1995c), and other localities mentioned in the text. TVZ, Taupo Volcanic Zone; CVZ, Coromandel Volcanic Zone (Skinner 1986). Rhyolitic volcanic centres: 1, Tuhua; 2, Okataina; 3, Rotorua; 4, Kapenga; 5, Reporoa; 6, Mangakino; 7, Maroa; 8, Whakamaru; 9, Taupo. Named volcanoes: H, Haroharo; T, Tarawera; To, Tongariro; N, Ngauruhoe; R, Ruapehu. VF, volcanic field; VC, volcanic centre.

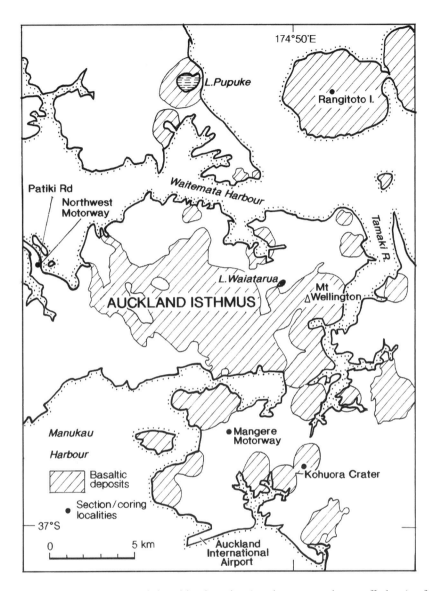

Fig. 2. Auckland Isthmus and basaltic deposits (scoriaceous tephra, tuff, lava) of the Auckland Volcanic Field (after Kermode 1992), and locations of sites referred to in the text. The map encompasses the majority of the Auckland urban area, which has a population of c. 1 million.

basaltic eruptions and, although most hazards are probably confined to a radius of 1-4 km, the concentration of population, buildings, transport and communication systems, and economic activity, together with uncertainties in locating the next eruptive centre, mean risk levels are high (Dibble *et al.* 1985; Cassidy *et al.* 1986; Houghton *et al.* 1988; Kermode 1992; Smith & Allen 1993; Allen & Smith 1994; Johnson *et al.* 1997). The main hazard types are destructive pyroclastic surges and small blasts including phreatomagmatic activity, fall of wet or dry tephra (i.e.

unconsolidated ash, lapilli, bombs or blocks), and aa and pahoehoe lava flows (Allen & Smith 1994; Johnson *et al.* 1997). The associated tsunami hazard in the Auckland region is comparatively minor — a recent assessment by de Lange & Hull (1994) concluded that the greatest threats are from earthquake-generated events or far-field volcanogenic tsunami from the west coast of South America.

However, the City of Auckland also faces a quite different volcanic hazard, namely that arising from deposits originating from distant rhyolitic and andesitic volcanic centres in the central North Island (CNI) (Fig. 1). These centres have been active during most of the Quaternary and are characterized by relatively high-magnitude, high-frequency eruptions. In this paper, evidence is presented that demonstrates that CNI volcanism is likely to be hazardous to the City of Auckland, and preliminary work towards an improved assessment of these hazards is described. It is our contention that the threat of tephra fall in particular from these distant volcanic centres has not been adequately assessed, and hence we suggest lines of research that could be undertaken to assist with a re-evaluation of the full spectrum of volcanic hazards that the city and surrounding region faces.

Quaternary volcanism in the central North Island

It has been estimated that *c.* 25% of the world's eruptions in historical or prehistorical times with a volcanic explosivity index of 5 or more were from the CNI of New Zealand; moreover, this region contains the world's strongest concentration of youthful rhyolite volcanoes (Simkin & Siebert 1994). In CNI, andesitic activity started *c.* 2 Ma ago and was joined by voluminous rhyolitic (plus minor basaltic and dacitic) activity from at least *c.* 1.6 Ma (Wilson *et al.* 1995*c*). This silicic volcanism has produced a complex of volcanic centres and covered much of the CNI with extensive plateaux of ignimbrites (deposits from pyroclastic flows) and mantling tephra fall deposits which together have bulk volumes *c.* 15 000-20 000 km^3 (Healy 1992; Houghton *et al.* 1995; Wilson *et al.* 1995*c*). The most extensive pyroclastic deposits are derived from rhyolitic caldera volcanoes of the central Taupo Volcanic Zone (TVZ) (Fig. 1). Smaller volumes have been erupted from the andesitic centres, chiefly Egmont and Tongariro volcanic centres, and from the offshore peralkaline Tuhua Volcanic Centre (Mayor Island). TVZ activity has continued to the present. Beyond these source regions, numerous tephra layers have been identified from their stratigraphic position, field properties, mineralogy, and chemical composition, and the radiometric or geomagnetic ages of many Quaternary units are now known with reasonable certainty (e.g. see Froggatt & Lowe 1990; Shane & Froggatt 1991; Kohn *et al.* 1992; Stokes *et al.* 1992; Alloway *et al.* 1993, 1995; Wilson 1993; Shane 1994; Houghton *et al.* 1995; Donoghue & Neall 1996; Naish *et al.* 1996; Pillans *et al.* 1996; Shane *et al.* 1996).

We focus below on five major eruptive centres of CNI that have been active during the Late Quaternary, these representing the greatest potential threat to the City of Auckland today. The reader is referred to Blong (1984, 1996) and Chester (1993) for comprehensive overviews of the effects of volcanic eruptions, and to Dibble & Neall (1984), Dibble *et al.* (1985), Gregory & Watters (1986), Houghton *et al.* (1988), Ministry of Civil Defence (1995), Johnson *et al.* (1997), and Houghton (1998) for more specific accounts of the principal volcanic hazards of North Island.

Tuhua (Mayor Island)

Situated 30 km from the Coromandel/Bay of Plenty coast and 140 km southeast of Auckland, this isolated, peralkaline rhyolite volcano has a complex history of eruptions dating from *c.* 130 ka (Houghton *et al.* 1992). It exhibits a wide range of eruption styles and intensity, with the largest known event occurring *c.* 6.2 ka ago (Tuhua Tephra). This tephra is widespread on North Island (Hogg & McCraw 1983; Lowe 1988; Newnham & Lowe 1991), but only one other event (Te Paritu Tephra at *c.* 14.5 ka) is recorded on North Island (Lowe 1988; Wilson *et al.* 1995a). Three other Mayor Island-derived tephras, all pre-Holocene, have been recorded in sediments in marine cores northeast from the volcano (Pillans & Wright 1992). Activity on Mayor Island since the Tuhua Tephra was deposited seems to be limited to caldera infilling by lavas within the last few thousand years (Houghton *et al.* 1992). The potential hazards for Auckland are confined to ashfall, and are likely to occur only from powerful eruptions taking place during easterlies (Buck 1985).

Okataina

Okataina is the world's second most productive rhyolitic volcanic centre (Wilson 1993; Wilson *et al.* 1995c). Situated *c.* 200 km south-southeast of Auckland, it consists of two major volcanoes, Haroharo and Tarawera. Tarawera erupted 10 June AD 1886, producing the biggest and most destructive eruption in New Zealand in historical times (Walker *et al.* 1984). Nine other large eruptive episodes are recognized spanning the last *c.* 21 000 years (Nairn 1989; Froggatt & Lowe 1990; Lowe *et al.* 1998), while a number of earlier events additionally deposited ignimbrites that individually exceeded 100 km^3 in magma volume (Dibble *et al.* 1985; Nairn 1989, 1991). Eight macroscopic tephra fallout layers derived from the Okataina Volcanic Centre, ranging from 1-110 mm in thickness, have been identified in post-18 ka lake sediments and peats near Hamilton in the Waikato region, *c.* 80-130 km south of Auckland (Lowe 1988; Hodder *et al.* 1991; Newnham *et al.* 1995). The principal hazards for Auckland are likely to be tephra fall during southeasterlies and southerlies and, in extreme cases, pyroclastic surge and flow with co-ignimbritic ash.

Taupo

Situated *c.* 230 km south of Auckland, Taupo is the single-most frequently active and productive rhyolitic volcano on Earth (Wilson 1993). The products of 28 eruptions postdating the voluminous 22 600 years BP Kawakawa (Oruanui) eruption (which generated $\geqslant = 800$ km^3 of erupted material) have been described by Wilson (1993). All but seven of these were plinian in dispersal. The latest major eruption was the immensely powerful ultraplinian Taupo Tephra event (equivalent to Unit Y of Wilson 1993) *c.* 1850 years BP. Because there is a wide spectrum of eruptive volumes and repose intervals, the timing and size of any future eruption cannot be predicted easily. Taking a longer-term perspective, Wilson (1993) suggested that in terms of time and erupted volumes, the post-22.6 ka record at Taupo may be considered as eruptive 'noise' superimposed on the more uniform, longer-term activity in the central Taupo Volcanic Zone as a whole, where very large eruptions such as the Kawakawa occur on the average at more evenly spaced intervals of around one per 40-60 thousand years (see also Houghton et al. 1995; Wilson et al. 1995c). At least five macroscopic tephra fallout layers derived from the Taupo Volcanic Centre, ranging from 2-50 mm in thickness, have been deposited in post-18 ka lake sediments and peats in the Waikato region (Lowe 1988; Hodder et al. 1991; Newnham et al. 1995). As with Okataina, the principal hazards for distant localities are likely to be tephra fall deposition and, in extreme events, pyroclastic surge and flow with co-ignimbrite ash.

Tongariro

The principal sources of local (CNI) volcanic risk today are the subduction-related andesitic volcanoes of the Tongariro Volcanic Centre (Dibble et al. 1985), situated 260-280 km south of Auckland. Contemporary activity at Mt Ruapehu (2797 m above sea-level (asl)), a large stratovolcano, has resulted in tephra fall at distances > 100 km away in recent decades. The AD 1995 and 1996 eruptions of Ruapehu produced a cumulative volume of 0.06 km^3 of tephra that was deposited > 300 km from source over an area > 30 000 km^2 (Houghton et al. 1996), and resulted in the temporary closure of eleven airports throughout the North Island. At Auckland, the International Airport was closed overnight on three occasions for c. 12 hours in mid-1996 (19 June; 8 & 9 July), whilst an associated deterioration in air quality was recorded (G. Fisher, pers. comm. 1996). Mt Ngauruhoe (2290 m asl) is essentially a parasitic cone on Mt Tongariro (1968 m asl) that has erupted, on the average, every second year since AD 1839, the latest major event being in AD 1974-5 (Nairn & Self 1978). These recent eruptions of Ruapehu and Ngauruhoe have been on a relatively minor scale in comparison with much of their previous activity. The Late Quaternary history of the centre has been described by Hackett & Houghton (1986), Donoghue et al. (1995, 1997), Donoghue & Neall (1996), Cronin et al. (1996) and Nairn et al. (1998). Eleven post-12 ka tephra fallout layers from the Tongariro Volcanic Centre have so far been identified as medium-coarse ash layers 1–20 mm thick in lake sediments in the Waikato region (Lowe 1988). Thin macroscopic and microscopic ash layers from Tongariro are also recorded in sediments of both Lake Tutira and Kaipo peat bog, respectively c. 120 km to the east and 150 km northeast of source (Lowe & Hogg 1986; Eden et al. 1993; Eden & Froggatt 1996; Lowe et al. 1999a). The potential impacts for Auckland are likely to be limited to ash fall deposition during southerlies.

Egmont (Taranaki)

The Taranaki landscape of western North Island is dominated by the 2518 m-high subduction-related andesitic Mt Egmont volcano (also known as Mt Taranaki), situated 280 km south of Auckland. The youngest and most southerly volcano of the Taranaki volcanic succession, Egmont is the second highest mountain in the North Island and the largest andesitic stratovolcano in New Zealand. The post-130 000 year stratigraphic record indicates that ash eruption and lahar inundation are typical recurring features of Egmont volcano. Since 28 ka, at least 76 tephras (with volumes > 10^7 m^3) have been erupted intermittently (Alloway et al. 1995), with the latest definitive eruption of the volcano at c. AD 1755 or possibly later, near c.

Table 1. *Some effects of the accumulation of critical thicknesses of tephra*

Thickness	Effects on humans and livestock[a]	Effects on property and vegetation
1 mm	Little or none. Minor short-lived respiratory problems. Rapid recovery.	Airports closed. Minor disruption of some water supplies. Electronic and magnetic media affected
1 cm	Fish and insects killed. Little or no visibility. Infections of respiratory tract, asthma. Inflamed eyes.	Rusting of exposed metal. Transport immobilized. Electrical shorting and transformer fires. Stormwater and sewage systems may become inoperative. Food crops will require protection or post-harvest cleaning.
10 cm	Serious respiratory problems. Some human deaths possible. Bird life killed. Livestock will require imported feed.	Power blackouts. Crops destroyed or severely damaged. Trees stripped and branches broken. Roof collapse may occur.
1 m	Deaths and injuries due to building collapse. Livestock need to be moved or protected.	Building collapse. Vegetation fires. Trees destroyed. Total failure of all lifelines. Emergency services inoperative. Severe siltation of drainageways, ports and harbours.
10 m	Enormous loss of human life through building collapse and burns. Total loss of livestock.	Long-term loss of land use. Widespread building collapse. Some drainage ways, ports and harbours permanently altered.

[a] The effects are probably conservative especially at the 1 and 10 m levels [after Houghton *et al.* (1988) and Blong (1996)].

AD 1860 (Druce 1966; Neall 1972, 1992; Neall & Alloway 1986; Lees & Neall 1993; Price *et al.* 1999; Lowe *et al.* 1999*b*). Though not as voluminous as their more silicic counterparts, Egmont-sourced tephra beds are now recognized as more frequently erupted and more widely distributed across the North Island than first envisaged. Twelve post-12 ka tephra fall beds have so far been identified as medium-coarse ash layers 1-5 mm thick in lake sediments and peats of the Waikato region, *c.* 150-200 km north-northeast from source (Lowe 1988; Hodder *et al.* 1991; Newnham *et al.* 1995). Macroscopically visible Egmont tephras have also been identified in sediments of Lake Tutira (Eden *et al.* 1993; Eden & Froggatt 1996) and Kaipo bog (Lowe *et al.* 1999*b*). The principal threat to Auckland is ash-fall during southerlies and southwesterlies.

What are the likely threats for Auckland?

For the Auckland region, most of the volcanic hazards associated with these CNI centres do not pose a major threat. The most obvious hazards include ignimbrites, co-ignimbrite ash fall, lahars, and seismic activity. While devastating close to source, most of these hazards will have negligible physical impact on the Auckland region except for perhaps deposition from extreme but very infrequent ignimbritic events. The main volcanic hazards considered likely to have a direct physical impact upon Auckland on a relatively frequent basis would be fallout of silicic tephras and associated aerosols. The nature and extent of impact of tephra fall will be related to the primary thickness of the deposit (Table 1).

In turn, the thickness, as well as particle size and composition of tephric material deposited distally, and the distances and directions that tephras are transported, will be influenced by a number of volcanological and meteorological factors. Volcanological factors include the style of eruption, direction of blast and orientation of vent, physical and chemical properties of the source magma and the degree of magma–water interaction, height of eruption column, volume of erupted material, and whether or not pyroclastic flows are generated (Cas & Wright 1987). The efficacy of tephra fall dispersal will also depend upon wind direction, especially that of the jet stream in the stratosphere, as well as the dispersal characteristics of the fall deposits and synoptic meteorological conditions prevailing between the volcanic centre and the extremes of the tephra dispersal range. Duration of the eruptive episode and whether this is continuous or interrupted will also be important (Wilson 1993). Distal tephra thicknesses may be increased by co-ignimbritic ash fall due to elutriation from pyroclastic flows (Sparks & Walker 1977; Froggatt 1982; Fisher & Schmincke 1984) or by premature fallout due to particle aggregation or scavenging (Cas & Wright 1987). Overthickening of tephras in stratigraphic sequence may also occur because of reworking or by the addition of contemporaneous non-tephric material to the tephra unit. Remobilization of tephra-fall deposits following severe rainstorms also may lead to substantial overthickening in drainage channels adjacent to large catchments.

Assessment of volcanic hazards at distal localities should also consider the possible impacts of acid aerosols generated by volcanoes. For example, fluorine-rich gas and ash associated with the Laki fissure eruption of AD 1773 destroyed crops and livestock throughout Iceland and killed 24% of the human population by starvation (Simkin, 1994); social, ecological and meteorological impacts of volcanogenic gases and aerosols associated with this eruption elsewhere in northern Europe have been reported (Grattan et al. 1998; Brayshay & Grattan 1999). It has been suggested that the concentration of acid volatiles adsorbed on the surface of tephra particles in an eruption cloud may increase with time spent in transport and distance travelled (Oskarsson 1980; J. Grattan, pers. comm. 1995). Clearly, the possibility of volcanogenic 'acid rain' from CNI eruptions exacerbating Auckland's urban air pollution levels in certain meteorological conditions needs to be considered.

New Zealand's comparatively brief historical and documented era means that evaluation of these potential hazards requires an examination of Quaternary tephrostratigraphical records in the Auckland region. Although urban development across the Auckland Isthmus (Fig. 2) has destroyed or obscured much of the Quaternary geology, much valuable work can still be undertaken there to assist with assessment of the tephra hazard for Auckland from CNI eruptions. Three types of investigation could be pursued profitably:

1. extrapolation from tephra isopachs determined for localities intermediate between Auckland and the CNI volcanoes;
2. review of CNI tephras previously reported in the Auckland and Northland areas, coupled with field observations at new sites;
3. investigation of the impact of past CNI tephra deposition on the Auckland environment using appropriate palaeoecological techniques.

The results of preliminary work that demonstrates the potential of such investigations are reported here.

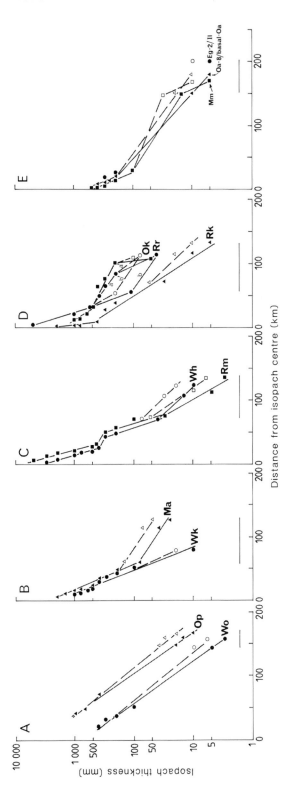

Fig. 3. Plots of tephra isopach thickness against distance from source for twelve post-18 ka CNI-derived tephras based on their distribution in the Waikato region (reprinted from Lowe 1988, fig. 13, p.157, with permission from SIR Publishing). Dashed curves join compaction-adjusted thicknesses (open symbols) for tephras recorded in Waikato lakes. Horizontal bars indicate region of Waikato lake core thickness measurements. Symbols for tephras: Wo, Whakaipo (Unit V of Wilson 1993); Op, Opepe (Unit E); Wk, Whakatane; Ma, Mamaku; Rm, Rotoma; Wh, Waiohau; Rr, Rotorua; Rk, Rerewhakaaitu; Ok, Okareka; Mm, Mangamate (?Waihohonu Lapilli: Donoghue et al. 1991); Oa-8/basal-Oa, Okupata; Eg-2/II, Englewood (Alloway et al. 1995).

Extrapolation from tephra isopachs

Lowe (1988) constructed provisional isopach maps for eleven post-15 ka CNI tephras recorded in lake sediment cores at 14 sites in the central Waikato region, near Hamilton, *c.* 100–200 km from the CNI volcanic centres and approximately midway between most of these and the City of Auckland (Fig. 1). Andesitic tephras (from Tongariro and Egmont) are more common in the south with their thicknesses decreasing markedly to the north and east. Lowe (1988) argued that the recorded tephra thicknesses probably underestimate equivalent dry-land thicknesses because of compaction or dissemination in the lake sediments and suggested that most of the 41 macroscopic tephras found in the Waikato lakes, erupted between *c.* 18 000 years BP and 1850 years BP, should also occur well beyond the Waikato area. Graphs showing isopach thicknesses plotted against distance from isopach centre suggest a near-linear semilog relationship indicating an exponential decrease in tephra thickness with increasing distance from source (Fig. 3). The change of slope or 'fine tail' in tephra distribution evident in some plots, due perhaps to co-ignimbritic ashfall or premature fallout as discussed above, or to the effects of strong winds, complicates attempts to extrapolate to greater distances using these plots. Nevertheless, they provide a basis for approximate estimates of the frequency with which CNI tephras have reached Auckland during the past *c.* 15 000 years and their possible local thicknesses.

Although the prospects of finding these tephras in the Auckland urban environment are comparatively limited (see below), it may also be useful to look for their occurrence beyond Auckland in the Northland region (Fig. 1). For example, the *c.* 665 years BP Kaharoa Tephra (from Okataina Volcanic Centre; Lowe *et al.* 1998), although not recorded in the western Waikato region (Lowe 1988; Hodder *et al.* 1991), occurs in Northland (Pullar *et al.* 1977; Newnham *et al.* 1998). Other CNI-derived rhyolitic tephras are recorded in Northland, including Taupo Tephra (1850 years BP) (Pullar *et al.* 1977), possibly Mamaku Tephra (7250 years BP), and four thin, uncorrelated ash-grade tephras older than *c.* 65 000 years BP (Lowe 1987). Kawakawa Tephra (22.6 ka) may also be present (Rankin 1973; Self 1983). Further, an Okataina-sourced tephra, the Rotoehu Ash (Froggatt & Lowe 1990), dated at 64 ± 4 ka (Wilson *et al.* 1992; Lowe & Hogg 1995), has been recorded as a *c.* 8-10 cm thick unit in lacustrine sediments at Lake Omapere *c.* 200 km north of Auckland (Fig. 1; Lowe 1987; Newnham 1999). On the basis of glass-shard major element chemistry and its distinctive cummingtonite-rich mineralogy (Froggatt & Lowe 1990), Rotoehu Ash has also been identified by the authors at Lake Rototuna *c.* 90 km northwest of Auckland. It also occurs as a discrete layer *c.* 14 cm thick in a motorway cutting just south of Auckland, and intermixed with other tephra materials in soil profiles near Pukekohe. Using isopach distributions mapped by previous workers, Lowe (1987) found a near linear log-normal relationship between Rotoehu Ash thickness and distance from source, which, together with the above evidence, indicated that this tephra must have been widely distributed across the Auckland Isthmus; however, it is only rarely observed as a definitive layer in Late Quaternary cover bed deposits.

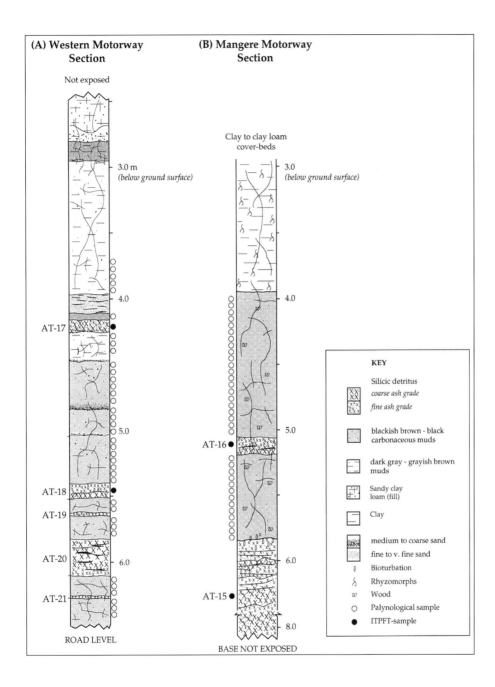

Fig. 4. Stratigraphic columns from (a) Northwestern Motorway (R11/595800 – 1 : 50 000 topographical map series grid reference of New Zealand Map Series 260), and (b) Mangere Motorway (R11/712689) sections showing uncorrelated silicic tephras and pollen and ITPFT (fission track) sampling positions.

Field surveys of silicic tephras in the Auckland region

Sequences of Quaternary rhyolitic tephra deposits, mostly reworked into estuarine deposits, and exposed at coastal localities in the Auckland region, have been described by Moore (1991). Some primary tephra fall deposits, up to 0.6 m thick, as well as rhyolitic ignimbrite units up to 9 m thick (e.g. distal ignimbrites at Port Waikato: Nelson et al. 1989; Kidnappers Ignimbrite, Auckland City: Wilson et al. 1995b), have been recorded. Deposits of reworked silicic detritus, up to 15 m thick, form a major component of the Quaternary subsurface deposits of the Auckland region. Major element glass compositions of these deposits suggest probable source areas in the Taupo Volcanic Zone, although the Coromandel Volcanic Zone, and potentially a much older age, cannot be ruled out (Nelson et al. 1989; Newnham & Grant-Mackie 1993). Clearly, more work needs to be done on the geochemical fingerprinting of these pyroclastic deposits, their stratigraphic relationships, and the impact their deposition might have had on the local environment. Despite the urban environment (and sometimes because of it) there are still opportunities to undertake this work.

For example, sections of rhyolitic tephras interbedded with organic carbonaceous sediments, possibly of Early to Mid-Quaternary age, have recently been exposed by road excavations and in coastal exposures at numerous localities in Auckland including the Northwest Motorway and Mangere Motorway sections (Figs 2, 4; Alloway & Newnham 1995). Investigations in progress at these sites include geochemical characterization using the electron microprobe and isothermal plateau fission-track (ITPFT) dating of the tephras to facilitate their correlation and to date surface and subsurface Quaternary deposits throughout the Auckland region. The study of these surface sections has been augmented recently by the acquisition of two long sedimentary cores extracted from a near sea-level locality at Patiki Road in west Auckland (Fig. 2). The c. 55 m-long cores comprise pollen-rich carbonaceous and laminated sediments interbedded with 19 discrete silicic tephra layers, some up to 0.9 m thick. Preliminary ITPFT dating and geochemical analysis of some of these tephras indicate that the sequence extends back more than 1.2 Ma. Because of the near-continuous deposition, and the abundance of discrete tephra beds, the Patiki Road cores can be used as a stratigraphic reference sequence to which deposits at the Northwest and Mangere motorway sections, and at other exposures and cores in the Auckland region and beyond, can be correlated. In this way, a more comprehensive record of the nature and distribution of CNI tephra deposits can be developed.

Other tephra-preserving sites (e.g. Lake Waiatarua and Kohuora Crater, described below) provide further potential to document the record of more recent tephra deposition in the Auckland region, as well as allowing assessment of the environmental impacts of such deposits.

Investigating the environmental impact of past central North Island-derived tephra deposition

Geomorphic processes associated with local volcanic activity have also provided opportunities to investigate the impacts of Late Quaternary CNI volcanism in Auckland. Lava flows have in places blocked drainage routes resulting in the

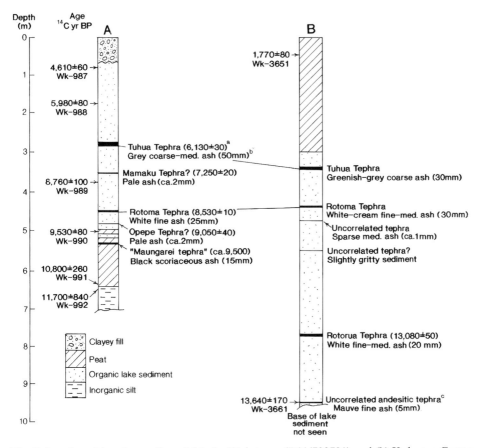

Fig. 5. Stratigraphic columns from (a) Lake Waiatarua (R11/732781) and (b) Kohuora Crater (R11/744675) showing CNI-derived tephras and radiocarbon dates. Wk- prefixes refer to the University of Waikato Radiocarbon Dating Laboratory number. The Kohuora core has been sampled for pollen analysis at 5 mm intervals above and below the tephra layers, and at c. 10 cm intervals for the rest of the sequence. Lake Waiatarua data after Newnham & Lowe (1991). [a] Tephra ages from Froggatt & Lowe (1990); [b] approximate thickness in parentheses; [c] possibly correlates with Rotoaira Lapilli, erupted from Mt Tongariro c. 13 700 years BP (Lowe 1988; Donoghue et al. 1995).

formation of lakes or swamps (Searle 1981; Newnham 1990; Newnham & Lowe 1991; Kermode 1992), while many craters have also been infilled with lake sediments and peat deposits. Where such sites have not been destroyed by urban development, they potentially provide continuous records of the local and regional environment since the originating eruptive episode and also may preserve tephra layers derived from subsequent CNI volcanic activity. Palynology has demonstrated the sensitivity of vegetation in northern New Zealand to climate change (e.g. Newnham & Lowe, 1991; Newnham et al. 1989, 1993), and so detailed palynological studies of the organic sediments to determine the local paleoenvironment, and any environmental impacts following tephra fall, are also being undertaken (see Giles et al. 1999).

Lake Waiatarua (Fig. 2), formed by blockage of drainage in a valley by lava flows, contains four macroscopic silicic tephra layers, between 2–50 mm in thickness and derived from three CNI sources (Fig. 5). These overlie the c. 9500 years BP basaltic Maungarei tephra erupted from nearby Mt Wellington (Newnham & Lowe 1991). The local environmental impact of the Mt Wellington eruption was significant, altering the local hydrology and geomorphology, modifying local vegetation during the eruptive episodes, and in controlling subsequent successional vegetation development. Possible impacts of CNI tephra deposition were not investigated at a fine scale at this site. However, a longer and potentially more detailed record is preserved in lacustrine and peaty sediments infilling Kohuora Crater, a tuff ring at Papatoetoe in South Auckland (Fig. 2). Six macroscopic silicic tephra layers (including at least one andesitic tephra tentatively attributed to the Tongariro Volcanic Centre) have so far been provisionally identified in a 9.5 m-long core from Kohuora (Fig. 5). Work in progress at this site involves further coring, radiocarbon dating, fine-resolution analysis of pollen, and geochemical analyses to determine the likely environmental impacts of the CNI tephra fall deposits. Preliminary results from radiocarbon dating indicate that the top of this sequence is younger than c. 1800 years BP and that the base is around 13 700 years BP. However, because the organic sediments at Kohuora Crater extend below 9.5 m depth, this site also presents an opportunity to reconstruct a detailed and continuous environmental history of Auckland extending from near present to well before the last glacial maximum. The chronology to be established at this site will provide a minimum age for the Kohuora eruption. Two radiocarbon dates, $29\,700 \pm 2000$ (NZ488) and $29\,000 \pm 700$ (NZ540) [laboratory number of the New Zealand Radiocarbon (now Rafter) Dating Laboratory, Lower Hutt], both on wood buried in tuff at this volcano (Grant-Taylor & Rafter 1971), require verification in view of the unknown 'inbuilt' age of the wood, the large standard deviations, and the known difficulties of contamination and age reliability towards the upper limits of the method's range.

At the Patiki Road site, preliminary pollen analysis of the long cores indicates that these have good potential for the palynological reconstruction of past climate change. Similar potential is evident at the adjunct Northwest and Mangere motorway sections. This palaeoenvironmental work, together with that on the younger deposits at Kohuora Crater and other sites, will be incorporated into a detailed interlocking and precise time-stratigraphic framework using tephrochronology in conjunction with ITPFT and radiocarbon dating and magnetostratigraphic techniques.

Conclusions

The latest eruption of Mt Ruapehu, commencing 17 June 1996, demonstrated that the assessment of Auckland's volcanic hazards must consider not just the threat from local basaltic volcanism, but also that from tephras erupted and dispersed from the distant CNI volcanoes. Although this recent activity is minor in comparison with earlier eruptions from the central North Island, there are significant risks to the City of Auckland because of its concentration of population, infrastructure and economic activity.

Preliminary research indicates that numerous silicic tephras have been deposited in the Auckland region during the Quaternary. Thicknesses of these primary tephra fall deposits, both rhyolitic and andesitic, range from c. 1 mm to $\geqslant = 0.6$ m. Potential effects of tephra deposition of this scale range from short-lived respiratory problems, closure of airspace, disruption of electronic and magnetic media, and possible disruption of water supplies and sewage and stormwater systems to serious respiratory problems, widespread injuries and death due to building collapse, power blackouts, fires, immobilization of transport and all emergency services, and likely destruction of trees, crops and livestock. Rhyolitic ignimbrites up to 9 m thick are also present in the Auckland region — they pose a very infrequent but extreme hazard that would result in enormous loss of human life through building collapse and burns. Volcanogenic 'acid rain' accompanying CNI eruptions is likely to be an additional hazard that would exacerbate Auckland's urban air pollution levels, and hence possible respiratory and other problems, in certain meteorological conditions.

Further research by the authors aims to geochemically characterize and radiometrically date tephra units from key sections and cores so that they may be more readily correlated to other tephra units occurring throughout the Auckland region and elsewhere. In addition, many of the silicic tephras are associated with organic muds and peats which provide opportunities to undertake palaeoecological and sedimentological investigations into vegetation and landscape response to tephra deposition during the Quaternary.

This research demonstrates the necessity to reappraise the volcanic hazard assessments for the Auckland region because the threat of inundation by tephras and aerosols from distal CNI sources has not yet been considered in any regional or local authority plan. Together with complementary work on the local Auckland Volcanic Field (e.g. Johnson *et al.* 1997), these investigations will lead to better understanding of the evolution of the Auckland landscape, its natural history, and the impacts of Quaternary volcanism.

Lessons from the Auckland study may be applicable to many large cities elsewhere that are not currently considered at risk from volcanism. Our approach recognizes the need to integrate risk and volcanic hazards' assessments, and to develop mitigation strategies for specific localities that may be vulnerable to a variety of hazards either from proximal or distant volcanic sources.

We are grateful to the British Council (Wellington, New Zealand) for funding towards the travel costs associated with this project through a Higher Education Link between Waikato and Plymouth universities. BVA and RMN acknowledge the financial support of the Marsden Fund, Royal Society of New Zealand (96-UOA-ESA-0010); DJL and BVA acknowledge financial support of the Auckland Regional Council (Environment), facilitated by Michelle Daly. Peter de Lange (Department of Conservation), Teresa Giles, and Calvin Mora assisted with fieldwork at Kohuora Crater. Russell Blong (Macquarie University) is especially thanked for useful comments on the paper. The paper is an output from the HITE ('Hazards and impacts of tephra eruptions') Working Group of the Commission on Tephrochronology and Volcanology, International Union for Quaternary Research.

References

ALLEN, S. R. & SMITH, I. E. M. 1994. Eruption styles and volcanic hazard in the Auckland Volcanic Field, New Zealand. *Geoscience Reports of Shizuoka University*, **20**, 5–14.

ALLOWAY, B. V. & NEWNHAM, R. M. 1995. A preliminary assessment of the threat posed by distal silicic volcanism based on the Middle–Late Quaternary tephrostratigraphic record. *Geological Society of New Zealand Miscellaneous Publication*, **81B**, 73–83.

——, NEALL, V. E. & VUCETICH, C. G. 1995. Late Quaternary (post *c.* 28,000 year B.P.) tephrostratigraphy of northeast and central Taranaki, New Zealand. *Journal of the Royal*

Society of New Zealand, **25**, 385–458.

——, PILLANS, B. J., SANDHU, A. S. & WESTGATE, J. A. 1993. Revision of the marine chronology in the Wanganui Basin, New Zealand, based on the isothermal plateau fission-track dating of tephra horizons. *Sedimentary Geology*, **82**, 299–310.

BLONG, R. J. 1984. *Volcanic hazards*. Academic Press, Sydney.

—— 1996. Volcanic hazard risk assessment. *In*: SCARPA, R. & TILLING, R. (eds) *Monitoring and Mitigation of Volcanic Hazards*. Springer-Verlag, Berlin, 675–698.

BRAYSHAY, M. & GRATTAN, J. 1999. Environmental and social responses in Europe to the 1783 Laki fissure eruption of the Laki fissure volcano in Iceland: a consideration of contemporary documentary evidence. *This volume*.

BROTHERS, R. N. & GOLSON, J. 1959. Geological and archaeological interpretation of a section in Rangitoto Ash on Motutapu Island, Auckland. *New Zealand Journal of Geology and Geophysics*, **2**, 569–577.

BUCK, M. D. 1985. An assessment of volcanic risk on and from Mayor Island, New Zealand. *New Zealand Journal of Geology and Geophysics*, **28**, 283–298.

CAS, R. A. F. & WRIGHT, J. F. 1987. *Volcanic Successions*. Allen & Unwin, London.

CASSIDY, J., LOCKE, C. A. & SMITH, I. E. M. 1986. Volcanic hazard in the Auckland region. *New Zealand Geological Survey Record*, **10**, 60–64.

CHESTER, D. K. 1993. *Volcanoes and society*. Edward Arnold, London.

CRONIN, S. J., NEALL, V. E. & PALMER, A. S. 1996. Geological history of the north-eastern ring plain of Ruapehu volcano, New Zealand. *Quaternary International*, **34–36**, 21–28.

DE LANGE, W. P. & HULL, A. G. 1994. *Tsunami hazard for the Auckland region - a report for the Auckland Regional Council*. Department of Earth Sciences, University of Waikato & Institute of Geological and Nuclear Sciences Limited.

DIBBLE, R. R & NEALL, V. E. 1984. Volcanic hazards in New Zealand. *In*: SPEDEN, I. G. & CROZIER, M. J. (eds) *Natural Hazards in New Zealand*. New Zealand National Commission for UNESCO, Wellington, 332–374

——, NAIRN, I. A. & NEALL, V. E. 1985. Volcanic hazards of North Island, New Zealand – overview. *Journal of Geodynamics*, **3**, 369–396.

DONOGHUE, S. L. & NEALL, V. E. 1996. Tephrostratigraphic studies at Tongariro Volcanic Centre, New Zealand: an overview. *Quaternary International*, **34–36**, 13–20.

——, —— & PALMER, A. S. 1995. Stratigraphy and chronology of late Quaternary andesitic tephra deposits, Tongariro Volcanic Centre, New Zealand. *Journal of the Royal Society of New Zealand*, **25**, 115–206.

——, ——, —— & STEWART, R. B. 1997. The volcanic history of Ruapehu during the past 2 millennia based on the record of Tufa Trig tephras. *Bulletin of Volcanology*, **59**, 136–146.

——, STEWART, R. B. & PALMER, A. S. 1991. Morphology and chemistry of olivine phenocrysts of Mangamate Tephra, Tongariro Volcanic Centre, New Zealand. *Journal of the Royal Society of New Zealand*, **21**, 225–236.

DRUCE, A. P. 1966. Tree ring dating of recent volcanic ash and lapilli, Mt. Egmont. *New Zealand Journal of Botany*, **4**, 3–41.

EDEN, D. N. & FROGGATT, P. C. 1996. A 6500-year-old history of tephra deposition recorded in the sediments of Lake Tutira, eastern North Island, New Zealand. *Quaternary International*, **34–36**, 55–64.

——, ——, TRUSTRUM, N. A. & PAGE, M. J. 1993. A multiple-source Holocene tephra sequence from Lake Tutira, Hawke's Bay, New Zealand. *New Zealand Journal of Geology and Geophysics*, **36**, 233–242.

FISHER, R. V. & SCHMINCKE, H. U. 1984. *Pyroclastic rocks*. Springer-Verlag, Berlin.

FROGGATT, P. C. 1982. Review of methods of estimating rhyolitic tephra volumes; applications to the Taupo Volcanic Zone, New Zealand. *Journal of Volcanology and Geothermal Research*, **14**, 301–318.

—— & LOWE, D. J. 1990. A review of late Quaternary silicic and some other tephra formations from New Zealand: their stratigraphy, nomenclature, distribution, volume, and age. *New Zealand Journal of Geology and Geophysics*, **33**, 89–109.

GILES, T. M., NEWNHAM, R. M., LOWE, D. J. & MUNRO, A. J. 1999. Impact of tephra fall and environmental change: a 1000 year record from Matakana Island, Bay of Plenty, North

Island, New Zealand. *This volume*.
GRANT-TAYLOR, T. L. & RAFTER, T. A. 1971. New Zealand radiocarbon age measurements – 6. *New Zealand Journal of Geology and Geophysics*, **14**, 364–402.
GRATTAN, J., BRAYSHAY, M. & SUELLER, J. 1998. Modelling the distal impacts from gases emitted during the eruption of Italian and Icelandic volcanoes in 1783. *Quaternaire*, **9**, 25–35.
GREGORY, J. G. & WATTERS, W. A. (eds) 1986. Volcanic hazards assessment in New Zealand. *New Zealand Geological Survey Record*, **10**, 1–104.
HACKETT, W. R. & HOUGHTON, B. F. 1986. Active composite volcanoes of Taupo Volcanic Zone. *New Zealand Geological Survey Record*, **11**, 61–114.
HEALY, J. 1992. Central Volcanic Region. *In*: SOONS, J. M. & SELBY, M. J. (eds) *Landforms of New Zealand* (2nd edn). Longman Paul, Auckland, 256–286.
HODDER, A. P. W., DE LANGE, P. J. & LOWE, D. J. 1991. Dissolution and depletion of ferromagnesian minerals from Holocene tephra layers in an acid bog, New Zealand, and implications for tephra correlation. *Journal of Quaternary Science*, **6**, 195–208.
HOGG, A. G. & MCCRAW, J. D. 1983. Late Quaternay tephras of the Coromandel Peninsula, North Island, New Zealand: A mixed peralkaline and calkalkaline tephra sequence. *New Zealand Journal of Geology and Geophysics*, **26**, 163–187.
HOUGHTON, B. F. 1998. Blow up! – volcanoes. *In*: HICKS, G. & CAMPBELL, H. (eds) *Awesome Forces – the Natural Hazards That Threaten in New Zealand*. Te Papa Press, 18–43.
——, LATTER, J. H. & FROGGATT, P. C. 1988. Volcanic hazards in New Zealand. *Geological Society of New Zealand Miscellaneous Publication*, **41d**, 1–54.
——, WEAVER, S. D., WILSON, C. J. N. & LANPHERE, M. A. 1992. Evolution of a Quaternary peralkaline volcano: Mayor Island, New Zealand. *Journal of Volcanology and Geothermal Research*, **51**, 217–236.
——, WILSON, C. J. N., MCWILLIAMS, M. O., LANPHERE, M. A., WEAVER, S. D., BRIGGS, R. M. & PRINGLE, M. S. 1995. Chronology and dynamics of a large silicic magmatic system: central Taupo Volcanic Zone. *Geology*, **23**, 13–16.
——, NEALL, V. E. & JOHNSON, D. 1996. *Eruption! Mount Ruapehu awakes*. Penguin Books, Auckland.
JOHNSON, D. M., NAIRN, I. A., THORDARSON, T. & DALY, M. 1997. Volcanic impact assessment for the Auckland Volcanic Field. *Auckland Regional Council Technical Publication*, **79**.
KERMODE, L. O. 1992. *Geology of the Auckland urban area*. 1:50 000. Institute of Geological & Nuclear Sciences Geological Map, **2**. Lower Hutt, New Zealand.
KOHN, B. P., PILLANS, B. J. & MCGLONE, M. S. 1992. Zircon fission track age for middle Pleistocene Rangitawa Tephra, New Zealand: stratigraphic and paleoclimatic significance. *Palaeogeography, Palaeoclimatology, Palaeoecology*, **95**, 73–94.
LEES, C. M. & NEALL, V. E. 1993. Vegetation response to volcanic eruptions on Egmont volcano, New Zealand, during the last 1500 years. *Journal of the Royal Society of New Zealand*, **23**, 91–127.
LOWE, D. J. 1987. *Studies on late Quaternary tephras in the Waikato and other regions in northern New Zealand, based on distal deposits in lake sediments and peats*. DPhil thesis, Library, University of Waikato, Hamilton.
—— 1988. Stratigraphy, age, composition, and correlation of late Quaternary tephras interbedded with organic sediments in Waikato lakes, North Island, New Zealand. *New Zealand Journal of Geology and Geophysics*, **31**, 125–165.
—— & HOGG, A. G. 1986. Tephrostratigraphy and chronology of the Kaipo Lagoon, an 11,500 year-old montane peat bog in Urewera National Park, New Zealand. *Journal of the Royal Society of New Zealand*, **16**, 25–41.
—— & —— 1995. Age of the Rotoehu Ash – comment. *New Zealand Journal of Geology and Geophysics*, **38**, 399–402.
——, NEWNHAM, R. M. & WARD, C. M. 1999a. Stratigraphy and chronology of a sequence of multi-sourced silicic tephras in a 15,000 ^{14}C yr BP montane peat bog in eastern North Island, New Zealand. *New Zealand Journal of Geology and Geophysics*, in press.
——, ——, MCFADGEN, B. F. & HIGHAM, T. F. G. 1999b. Tephras and archaeology in New Zealand and the timing of earliest Polynesian settlement. *Journal of Archaeological Science*, in press.

———, McFadgen, B. F., Higham, T. F. G., Hogg, A. G., Froggatt, P. C. & Nairn, I. A. 1998. Radiocarbon age of the Kaharoa Tephra, a key marker for late-Holocene stratigraphy and archaeology in New Zealand. *The Holocene*, **8**, 487–495.

Ministry of Civil Defence 1995. Volcanic hazards in New Zealand. *Tephra*, **14(2)**, 1–33.

Moore, C. L. 1991. The distal record of explosive rhyolitic volcanism: an example from Auckland, New Zealand. *Sedimentary Geology*, **74**, 25–38.

Nairn, I. A. 1989. *Sheet V16 AC – Mount Tarawera. Geological Map of New Zealand 1 : 50 000*. New Zealand Geological Survey, Lower Hutt.

——— 1991. Volcanic hazards at Okataina Volcanic Centre. *Ministry of Civil Defence, Volcanic Hazards Information Series*, **2**, 1–29.

——— & Self, S. 1978. Explosive eruptions and pyroclastic avalanches from Ngauruhoe in February, 1975. *Journal of Volcanology and Geothermal Research*, **3**, 39–60.

———, Kobayashi, T. & Nagakawa, M. 1998. The ~10 ka multiple vent pyroclastic eruption sequence of Tongariro Volcanic Centre, Taupo Volcanic Zone, New Zealand: Part 1. Eruptive processes during regional extension. *Journal of Volcanology and Geothermal Research*, **86**, 19–44.

Naish, T. R., Kamp, P. J. J., Alloway, B. V., Pillans, B. J., Wilson, G. S. & Westgate, J. A. 1996. Integrated tephrochronology and magnetostratigraphy for cyclothemic marine strata, Wanganui Basin: implications for redefining the Plio-Pleistocene boundary in New Zealand. *Quaternary International*, **34–36**, 29–48.

Neall, V. E. 1972. Tephrochronology and tephrostratigraphy of western Taranaki (N108–109), New Zealand. *New Zealand Journal of Geology and Geophysics*, **15**, 507–557.

——— 1992. Landforms of Taranaki and the Wanganui lowlands. *In*: Soons, J. M. & Selby, M. J. (eds) *Landforms of New Zealand* (2nd edn). Longman Paul, Auckland, 287–307.

——— & Alloway, B. V. 1986. Quaternary volcaniclastics and volcanic hazards of Taranaki. *New Zealand Geological Survey Record*, **12**, 101–137.

Nelson, C. S., Kamp, P. J. J. & Mildenhall, D. C. 1989. Late Pliocene distal silicic ignimbrites, Port Waikato, New Zealand: implications for volcanism, tectonics, and sea-level changes in South Auckland. *New Zealand Journal of Geology and Geophysics*, **32**, 357–370.

Newnham, R. M. 1990. *Late Quaternary palynological investigations into the history of vegetation and climate in northern New Zealand*. PhD thesis, Library, University of Auckland, Auckland.

——— 1999. Environmental change in Northland, New Zealand during the Last Glacial and Holocene. *Quaternary International*, in press.

——— & Grant-Mackie, J. A. 1993. Palynology of a peat layer interbedded with rhyolitic tephra layers at Bucklands Beach, Auckland: a preliminary investigation. *Tane*, **34**, 133–140.

——— & Lowe, D. J. 1991. Holocene vegetation and volcanic activity of the Auckland Isthmus, New Zealand. *Journal of Quaternary Science*, **6**, 177–193.

———, de Lange, P. J. & Lowe, D. J. 1995. Holocene vegetation, climate and history of a raised bog complex, northern New Zealand, based on palynology, plant macrofossils and tephrochronology. *The Holocene*, **5**, 267–282.

———, Lowe, D. J. & Green, J. D. 1989. Palynology, vegetation and climate of the Waikato lowlands, North Island, New Zealand, since c. 18,000 years ago. *Journal of the Royal Society of New Zealand*, **19**, 127–150.

———, ———, McGlone, M. S., Wilmshurst, J. M. & Higham, T. F. G. 1998. The Kaharoa Tephra as a critical datumn for earliest human impacts in northern New Zealand. *Journal of Archaeological Science*, **25**, 533–544.

———, Ogden, J. & Mildenhall, D. C. 1993. Late Pleistocene vegetation history of the Far North of New Zealand. *Quaternary Research*, **39**, 361–372.

Nichol, R. 1992. The eruption history of Rangitoto: reappraisal of a small New Zealand myth. *Journal of the Royal Society of New Zealand*, **22**, 159–180.

Oskarsson, N. 1980. The interaction between volcanic gases and tephra: fluorine adhering to tephra of the 1970 Hekla eruption. *Journal of Volcanology and Geothermal Research*, **8**, 251–266.

Pillans, B. J. & Wright, I. 1992. Late Quaternary tephrostratigraphy from the southern

Havre Trough-Bay of Plenty, northeast New Zealand. *New Zealand Journal of Geology and Geophysics*, **35**, 129–143.

——, KOHN, B. P., BERGER, G. W., FROGGATT, P. C., DULLER, G., ALLOWAY, B. V. & HESSE, P. P. 1996. Multi-method dating comparison from Mid-Pleistocene Rangitawa Tephra, New Zealand. *Quaternary Science Reviews*, **15**, 1–14.

PRICE, R. C., STEWART, R. B., WOODHEAD, J. D. & SMITH, I. E. M. 1999. Petrogenesis of high-K arc margins: evidence from Egmont volcano, North Island, New Zealand. *Journal of Petrology*, **40**, 167–197.

PULLAR, W. A., KOHN, B. P. & COX, J. E. 1977. Air-fall Kaharoa Ash and Taupo Pumice, and sea-rafted Loisels Pumice, Taupo Pumice, and Leigh Pumice in northern and eastern parts of the North Island, New Zealand. *New Zealand Journal of Geology and Geophysics*, **20**, 697–717.

RANKIN, P. C. 1973. Correlation of volcanic glasses in tephras and soils using micro-element compositions. *New Zealand Journal of Geology and Geophysics*, **16**, 637–641.

SEARLE, E. J. 1981. *City of volcanoes: a geology of Auckland*. Longman Paul, Auckland.

SELF, S. 1983. Large-scale phreatomagmatic silicic volcanism: a case study from New Zealand. *Journal of Volcanology and Geothermal Research*, **17**, 433–469.

SHANE, P. A. R. 1994. A widespread, early Pleistocene tephra (Potaka Tephra, 1 Ma) in New Zealand: character, distribution, and implications. *New Zealand Journal of Geology and Geophysics*, **37**, 25–36.

—— & FROGGATT, P. C. 1991. Glass chemistry, paleomagnetism, and correlation of middle Pleistocene tuffs in southern North Island, New Zealand, and Western Pacific. *New Zealand Journal of Geology and Geophysics*, **34**, 203–211.

——, ALLOWAY, B. V., BLACK, T. & WESTGATE, J. A. 1996. Isothermal plateau fission-track ages of tephra beds in an Early–Middle Pleistocene marine and terrestrial sequence, Cape Kidnappers, New Zealand. *Quaternary International*, **34–36**, 49–54.

SIMKIN, T. 1994. Distant effects of volcanism — how big and how often? *Science*, **264**, 913–914.

—— & SIEBERT, L. 1994. *Volcanoes of the world* (2nd edn). Smithsonian Institution, Geosciences Press, Tucson.

SKINNER, D. N. B. 1986. Neogene volcanism of the Hauraki Volcanic Region. *The Royal Society of New Zealand Bulletin*, **23**, 21–47.

SMITH, I. E. M. & ALLEN, S. R. 1993. Volcanic hazards at the Auckland Volcanic Field. *Ministry of Civil Defence, Volcanic Hazards Information Series*, **5**, 1–34.

SPARKS, R. S. J. & WALKER, G. P. L. 1977. The significance of vitric-enriched air-fall ashes associated with crystal-enriched ignimbrites. *Journal of Volcanology and Geothermal Research*, **2**, 329–341.

STOKES, S., LOWE, D. J. & FROGGATT, P. C. 1992. Discriminant function analysis and correlation of late Quaternary rhyolitic tephra deposits from Taupo and Okataina volcanoes, New Zealand, using glass shard major element composition. *Quaternary International*, **13–14**, 103–117.

WALKER, G. P. L., SELF, S. & WILSON, L. 1984. Tarawera 1886, New Zealand — a basaltic plinian fissure eruption. *Journal of Volcanology and Geothermal Research*, **21**, 61–78.

WILSON, C. J. N. 1993. Stratigraphy, chronology, styles and dynamics of late Quaternary eruptions from Taupo volcano, New Zealand. *Philosophical Transactions of the Royal Society of London*, **A343**, 205–306.

——, HOUGHTON, B. F., LANPHERE, M. A. & WEAVER, S. D. 1992. A new radiometric age estimate for the Rotoehu Ash from Mayor Island volcano, New Zealand. *New Zealand Journal of Geology and Geophysics*, **35**, 371–374.

——, ——, PILLANS, B. J. & WEAVER, S. D. 1995a. Taupo Volcanic Zone calc-alkaline tephras in the peralkaline Mayor Island volcano, New Zealand. *Journal of Volcanology and Geothermal Research*, **69**, 303–311.

——, ——, KAMP, P. J. J. & MCWILLIAMS, M. O. 1995b. An exceptionally widespread ignimbrite with implications for pyroclastic flow emplacement. *Nature*, **378**, 605–607.

——, ——, MCWILLIAMS, M. O., LANPHERE, M. A., WEAVER, S. D. & BRIGGS, R. M. 1995c. Volcanic and structural evolution of Taupo Volcanic Zone, New Zealand: a review. *Journal of Volcanology and Geothermal Research*, **68**, 1–28.

The role of tephrachronology in recognizing synchronous caldera-forming events at the Quaternary volcanoes Longonot and Suswa, south Kenya Rift

STUART C. SCOTT[1] and IAN P. SKILLING[2]

[1] *Department of Geological Sciences, University of Plymouth, Drake Circus, Plymouth PL4 8AA, UK*
[2] *Department of Geology, Rhodes University, Box 94, Grahamstown, 6140 South Africa*

Abstract: The Quaternary volcanoes Longonot and Suswa, located in the south Kenya Rift, each have a caldera formed by incremental collapse accompanied by airfall, pyroclastic flow and pyroclastic surge activity. To constrain the relative ages of caldera formation at these two closely spaced centres, syn-caldera deposits are correlated using tephrachronology.

Volcanic products from Longonot and Suswa display contrasting Nb/Zr ratios which provide a unique signature for each centre. Early syn-caldera pyroclastics at Suswa contain airfall pumice beds which have Longonot Nb/Zr signatures. Variation of FeO(T) and Al_2O_3 with Zr is particularly effective at discriminating the various Longonot units in two component space, and has been combined with volcanological evidence to identify the equivalent Longonot deposits as either early syn-caldera or late syn-caldera. In both cases, the correlation indicates synchronous initiation of caldera formation. This suggests that caldera formation is a regional rather than local event in the southern part of the rift, highlighting the potential rôle of regional tectonics as a controlling mechanism.

Major alignments of eruptive vents, fissures and the caldera collapse centre at each volcano are parallel to rift faults, suggesting that rift floor tension fractures or normal faults are utilized as conduits for magma transport. Synchronous caldera formation at Longonot and Suswa may, therefore, be linked to lateral magma withdrawal along rift floor fractures and decompression of the shallow-level magma chambers during a period of rift floor extension.

The neighbouring Quaternary silicic caldera volcanoes Longonot and Suswa lie between 0°55′ S, 36°26′ E and 1°10′ S, 36°20′ E, at the southern end of a chain of eight Quaternary caldera complexes and two non-caldera Quaternary complexes in the central trough of the Gregory Rift, Kenya (Figs 1 & 2). Syn- and post-caldera pyroclastics at Longonot and Suswa are in a similar state of erosion, which suggests that caldera formation may have been broadly contemporaneous. If so, caldera formation may be a regional event in this part of the rift floor, highlighting the potential rôle of regional tectonics as a major control on volcano-structural events (Skilling 1993; Cambray *et al.* 1995).

In an attempt to provide a better constraint on the relative ages of caldera formation at these two centres, syn-caldera activity has been correlated by using

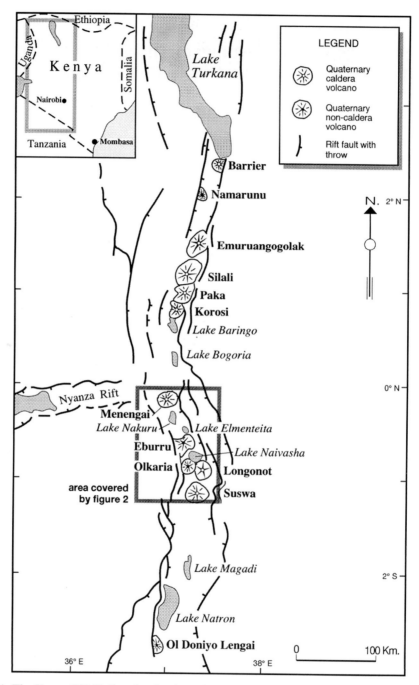

Fig. 1. The Kenya Rift Valley showing boundary faults and Quaternary–Recent volcanoes.

tephrachronology. This approach is based on the fact that, although syn- and post-caldera airfall pyroclastics from both centres were dispersed predominantly westwards, some of the Longonot airfall pumice was also dispersed southwestwards over Suswa, providing marker horizons within the Suswa syn-caldera succession.

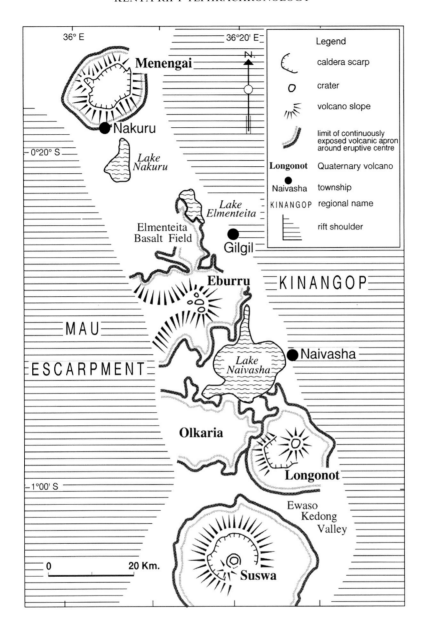

Fig. 2. Quaternary–Recent volcanoes of the south-central Kenya Rift.

Events at the time of caldera collapse

Longonot

The geology of Longonot has been described by Scott (1980) and Clarke *et al.* (1990). Stratigraphic nomenclature, lithology and structural events for the volcanic succession are summarized in Fig. 3.

Construction of an early 1200 m high, 25 km diameter composite volcano (Lt1 –

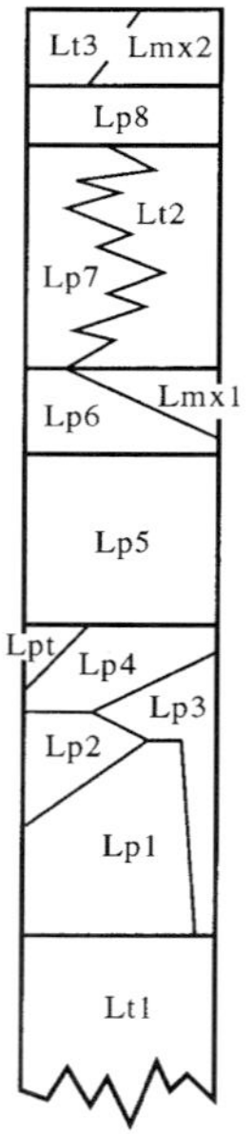

FORMATION	MEMBER	LITHOLOGY	EVENTS
Longonot Trachyte Formation (Lt)	Upper Trachyte Member (Lt3)	Trachyte lava	
Longonot Mixed Lava Formation (Lmx)	Upper Mixed-lava Member (Lmx2)	Mixed hawaiite - trachyte lava	
Akira Pumice Formation (Lpa)	Ash Member (Lp8)	Trachyte airfall ash	Summit crater formation
Longonot Trachyte Formation (Lt)	Lower Trachyte Member (Lt2)	Trachyte lava	Building of lava pile
Akira Pumice Formation (Lpa)	Upper Pumice Member (Lp7)	Trachyte airfall pumice	
Longonot Mixed Lava Formation (Lmx)	Lower Mixed-lava Member (Lmx1)	Mixed hawaiite - trachyte lava	
	Bedded Ash Member (Lp6)	Trachyte airfall ash	
Akira Pumice Formation (Lpa)	Lower Pumice Member (Lp5)	Trachyte airfall pumice	Post - caldera Plinian pyroclastic eruptions
Kedong Valley Tuff Formation (Lpk)	Tuff Cone Member (Lpt)	Breccia	
	Surge Member (Lp4)	Trachyte airfall pumice and surge	
	Upper Ignimbrite Member (Lp3)	Trachyte ignimbrites and airfall pumice	End of caldera formation
Akira Pumice Formation (Lpa)	Bedded Pumice Member (Lp2)	Trachyte airfall pumice	
Kedong Valley Tuff Formation (Lpk)	Lower Ignimbrite Member (Lp1)	Trachyte ignimbrites and airfall pumice	Start of caldera formation
Longonot Trachyte Formation (Lt)	Olongonot Volcanic Member (Lt1)	Trachyte lavas and pumice	Pre - caldera composite cone

Fig. 3. Stratigraphic nomenclature, lithology and volcano-structural events for the Longonot volcanic succession.

Olongonot Volcanic Formation) commenced <0.4 Ma ago. This was terminated by incremental collapse to form a c. 7.5 km diameter caldera, offset to the east of the original composite volcano summit. The caldera has a minimum volume of c. 26.5 km^3 (Scott 1980). Formation of this caldera was accompanied by pyroclastic activity which produced airfall ash, airfall pumice and ignimbrites.

The initial stage of caldera collapse was marked by the emplacement of an apron of five trachyte ignimbrites around the volcano accompanied by airfall pumice and ash (Lp1 – Lower Ignimbrite Member of the Kedong Valley Tuff Formation). Each of the five ignimbrite cooling units has a weathered top, suggesting significant time gaps between eruptions. This was followed by the deposition of a well-bedded trachyte airfall pumice unit from a pulsating Plinian eruption column, the bedding being defined by rapid changes in pumice clast size (Lp2 – Bedded Pumice Member of the Akira Pumice Formation). The final stage of collapse was marked by the emplacement of a second widespread trachyte ignimbrite (Lp3 – Upper Ignimbrite Member of the Kedong Valley Tuff Formation). Caldera formation promoted the interaction of groundwater with the magmatic system, culminating in a series of early post-caldera phreatomagmatic eruptions which deposited a sequence of trachyte pumice beds and base surge units (Lp4 – Surge Member of the Akira Pumice Formation). Post-caldera activity continued with a series of sporadic Plinian eruptions which deposited a succession of pumice lapilli beds (Lp5 – Lower Pumice Member of the Akira Pumice Formation), each bed of which is separated from the next by a palaeosol or oxidized horizon. The eruptions deposited a particularly thick but poorly consolidated mantle of pumice beds over the western caldera scarp. During deposition, a critical slope angle was reached on the scarp face resulting in slope failure and the present scalloped form of the western caldera rim.

Suswa

The geology of Suswa has been described by Johnson (1969) and by Skilling (1993). Stratigraphic nomenclature, lithology and structural events for the volcanic succession are summarized in Fig. 4.

Activity at Suswa began at <0.4 Ma ago with the construction of a 28 km diameter trachyte lava shield (S1 – Angat Kitet Formation). The presence of an overlying palaeosol and extensive dissection of the shield lavas demonstrate a period of quiescence following shield-building activity. Activity resumed with the initiation of incremental collapse of caldera 1, accompanied by a series of low volume pyroclastic eruptions from vents on an outer ring feeder zone, producing a deposit 15–20 m thick (S2 – Olgumi Formation). The succession consists of two trachyte globule ignimbrites, three carbonate–trachyte ignimbrites, eight trachyte pumice lapilli beds and a trachybasaltic ash (Fig. 5, section A). As caldera collapse continued, trachyte agglutinate flows were erupted from the outer ring feeder zone (S3 – Oloolwa Formation). In the final stages of caldera collapse, activity was confined to vents in the west of the caldera and along the Enkorika Fissure Zone on the northern flank. Activity in the west of the caldera was characterized by a sporadic series of pyroclastic eruptions, producing a succession of trachyte pumice lapilli beds, occasional agglutinate flows and a terminal dry surge deposit (S4 – Esinoni Formation) which mantle the western caldera rim. Palaeosols separate many individual beds (Fig. 5, section B). Low volume, trachyte lava flows and

FORMATION	LITHOLOGY	EVENTS
Eululu Formation (S8)	Phonolite lavas	Resurgence of 'Island Block'
Ol Doinyo Onyoke Formation (S7)	Phonolite lavas	Caldera 2 formation
Entarakua Formation (S6)	Phonolite lavas	
Enkorika Formaation (S5)	Trachyte lavas and domes	End of caldera 1 formation
Esinoni Formation (S4)	Trachyte airfall pumice and dry surges	
Oloolwa Formation (S3)	Trachyte agglutinate flows	
Olgumi Formaation (S2)	Trachyte ignimbrites, mixed carbonate-trachyte ignimbrites, trachyte airfall pumice	Start of caldera 1 formation
Angat Kitet Formation (S1)	Trachyte shield lavas	Pre - caldera shield

Fig. 4. Stratigraphic nomenclature, lithology and volcano-structural events for the Suswa volcanic succession.

domes (S5 – Enkorika Formation) were then erupted from the Enkorika Fissure Zone, the lavas flowing up to 1.5 km from the fissure. Caldera 1 has a diameter of $c.$ 12 km, covers an area of $c.$ 113 km^2 and has a minimum volume of $c.$ 22 km^3. Post-caldera 1 activity produced a sequence of phonolite lava flows and minor ash falls from vents on the floor of the caldera. The smaller, concentric, 5.5 km diameter caldera 2 formed on the infilled floor of caldera 1 during this period.

Comparison of eruptive activity during caldera formation at Longonot and Suswa

Although the Suswa caldera is slightly larger than that at Longonot, there are similar elements in the mechanism of collapse and the style of eruptive activity which accompanied caldera formation at these centres. Similarities include:

- incremental caldera collapse;
- a succession of ignimbrites and trachyte pumice lapilli beds associated with initiation of incremental collapse;
- a pyroclastic succession dominated by a series of pumice lapilli beds separated by palaeosols characterizing the late-caldera S4 formation on Suswa and the early post-caldera Lp4 and Lp5 members on Longonot.

These similarities suggest the possibility of a common mechanism controlling caldera formation at the neighbouring centres.

Tephrachronology

Methodology

The geochemical signature of Longonot-derived tephra units has been used to correlate activity at Longonot and Suswa. Longonot samples collected in 1973 and 1974 were analysed at Reading University. Major oxide abundances in glassy

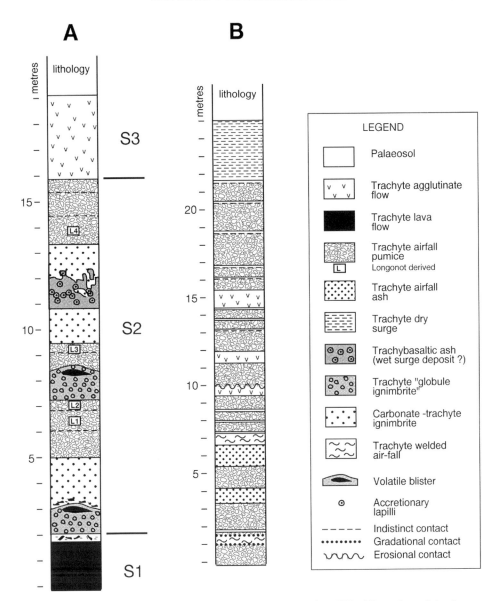

Fig. 5. (A) Composite section of the Suswa Olgumi Formation (S2); (B) section of the Suswa Esinoni Formation (S4) at UTM Grid Zone 37M, grid reference AJ997750.

samples were determined by wet chemical techniques whilst major oxide abundances in remaining samples and all trace-element abundances were determined using a Phillips PW 1410 X-ray fluorescence spectrometer. Longonot samples collected in 1986 and 1987, and all Suswa samples were analysed at Lancaster University. Major oxide and trace-element abundances were determined using a Phillips PW 1400 automatic X-ray fluorescence spectrometer.

Nine Longonot samples analysed at Reading University and covering a wide range of element concentrations, were re-analysed at Lancaster University to assess interlaboratory analytical precision. Good interlaboratory precision was obtained on all elements except Nb, Rb, Y and Zr. For each of these elements, the value of the Lancaster determination of abundance in each of the test samples needed to be multiplied by a constant factor to ensure compatibility with the Reading determination. These factors were as follows, with Standard Error estimates shown in brackets [ ] after the slope and intercept values :

Nb: (Lancaster result × 1.23 [0.05]) −0.32 [8.98] = Reading result
Rb: (Lancaster result × 1.19 [0.08]) −7.71 [10.18] = Reading result
Y: (Lancaster result × 1.23 [0.04]) −2.51 [3.36] = Reading result
Zr: (Lancaster result ×1.28 [0.03]) −1.85 [17.25] = Reading result

Application of these factors to the Lancaster determinations ensured compatibility of element determinations made at the two laboratories.

Chemical signatures for each volcano

The volcanic products at Longonot and Suswa can be uniquely characterized using the trace elements Nb and Zr. The abundances of these two high field strength elements (HFSE) in the volcanic rocks reflect magmatic values, these elements being immobile during processes of rapid crystallization, chilling to a glass and secondary hydration (Baker & Henage 1977; Jones 1981; Weaver et al. 1990). Nb and Zr also have similar and low distribution coefficients in the dominant alkali feldspar phenocryst phase characteristic of the peralkaline silicic magmas of Longonot and Suswa (Wörner et al. 1983; Mahood & Stimac 1990). This promotes their incompatible behaviour during processes involving crystal-liquid equilibria, such as feldspar fractionation. The abundance of Nb and Zr gradually increases in the residual magma as the magma fractionates, whilst the ratio of the two elements remains relatively constant. The value of this ratio is dictated by the ratio of the two elements in the original magma before fractionation commences. Small differences in the ratio of Nb and Zr in the parental magmas at individual volcanic centres will be preserved as fractionation proceeds.

Variations of Nb with Zr for the volcanics from Longonot and Suswa are presented in Fig. 6a which demonstrates a more tightly constrained trend, with a lower Nb/Zr ratio for the Longonot sample suite compared to that of Suswa. This contrast is particularly well revealed in Fig. 6b, which shows the variation of Nb/Zr with Zr. Fig. 6b demonstrates a relatively constant Nb/Zr ratio of about 0.26 with increasing Zr for the Longonot volcanics in contrast to higher, but generally decreasing Nb/Zr ratios for the Suswa volcanics. This decreasing ratio would suggest that, although each element has a bulk distribution coefficient $D < 1$, for the Suswa suite $D_{Nb} > D_{Zr}$.

Figure 6 also reveals Suswa volcanics which display a Longonot Nb/Zr signature. These samples are from four trachyte airfall pumice beds within the early syn-caldera Olgumi Formation (S2), the beds involved being denoted L1 to L4 on Fig. 5, section A. All these samples are derived from outcrops on the north and northeast flanks of Suswa which directly face the cone and caldera of Longonot situated 15 km to the northeast. These chemical and field characteristics suggest a Longonot source for the four airfall pumice beds. Derivation from the only other large centre active in

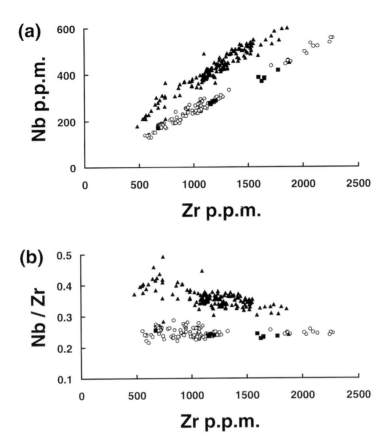

Fig. 6. (a) Variation of Nb with Zr for the volcanics from Longonot and Suswa. (b) Variation of Nb/Zr with Zr for the volcanics from Longonot and Suswa. Filled triangles – Suswa volcanics; open circles – Longonot volcanics; filled squares – Suswa S2 samples displaying a Longonot Nb/Zr signature.

the area, the Olkaria centre located to the north of Suswa (Fig. 2), seems unlikely as most of its trachyte airfall pumice was erupted very early in the history of the centre and dispersed westwards over the Mau Escarpment (Clarke et al. 1990). Comendite compositions dominate the later lava and pyroclastic succession of the Olkaria centre.

Correlation of Longonot and Suswa activity

The effectiveness of the Longonot-derived Suswa beds as marker horizons for correlating volcanic activity depends on establishing their equivalent stratigraphic position within the Longonot succession. Multivariate discriminant analysis, using chemical parameters to characterize individual stratigraphic members, leads to geologically unrealistic or ambiguous stratigraphic assignments for many of the unknown Longonot-derived Suswa samples. The variables Al_2O_3, FeO(T), Rb, Y, Zr, Nb, Ce and Pb, account for most of the chemical variation recorded in the Longonot volcanics. Assignments determined using this group of elements as variables, expressed as a percentage of the 11 Suswa samples being classified, are:

Lp1 – 27.3%; Lp5 – 54.5%; Lt2 – 18.2%

whilst assignments determined using the above elements plus Ba and Sr as variables are:

Lp1 – 27.4%; Lp3 – 18.3%; Lp5 – 9.2%; Lt2 – 45.1%

Assignments determined using all elements as variables are:

Lt1 – 9.1%; Lp5 – 54.5%; Lt2 – 27.3%; Lt3 – 9.1%

The Lt2 and Lt3 members are dominated by lava flows, although thin pumice horizons are found between the Lt2 lava flows. These individual, thin horizons could not, however, account for the thicknesses of Longonot-derived pumice beds found on Suswa. Geological reasons for the unacceptability of the Lp5 member as an equivalent of the Longonot-derived Suswa pumice are given below.

The unrealistic assignments are primarily the result of sample population overlaps in multicomponent space for each of the stratigraphic members. Unknown samples, assigned on an individual basis, often fall in overlap regions. Assignment as groups of samples constrained by known geological relationships, rather than assignment as individual samples, is more appropriate for discriminating the stratigraphic equivalence of unknown Longonot-derived Suswa samples taken from chemically zoned pumice beds. This approach, which cannot be accommodated in a discriminant analysis, is most effectively undertaken in two component space. Variation in concentration of the major oxides FeO(T) and Al_2O_3 with that of the incompatible trace element Zr, is particularly effective at discriminating the chemistry of individual Longonot stratigraphic members in two component space.

Figures 7 and 8 show plots of Zr against FeO(T) and Zr against Al_2O_3 for each of the Longonot stratigraphic units and for the Longonot-derived Suswa pumice. Symbols for each stratigraphic unit are shown on Fig. 7. Starting with the youngest units, Figs 7c and 8c show an overlap between the Longonot-derived Suswa pumice and the Longonot Lp5 pumice. However, the possibility that the Longonot Lp5 pumice is an equivalent of the Longonot-derived Suswa pumice is disregarded since:

(i) the spread of Longonot derived Suswa pumice does not in general coincide with the established trends of individual zoned Lp5 pumice beds;
(ii) at least two of the Longonot-derived Suswa pumices overlap the fractionation trend of the Longonot Lt2 lava pile;
(iii) two of the Longonot derived Suswa pumice samples show FeO(T) abundance significantly above the maximum abundance of 7.72 wt. % recorded in the Lp5 beds;
(iv) the Longonot-derived Suswa pumice contains 13–48 ppm Ba and 1–28 ppm Sr (Table 1) in contrast to the Longonot Lp5 pumice which contains <10 ppm Ba for all but the two youngest Lp5 samples (which record 11 ppm and 70 ppm Ba) and <5 ppm Sr (Table 2);
(v) prominent palaeosols and oxidised horizons separate individual beds within the Lp5 succession, attesting to periods of quiescence between eruptions. These are absent between adjacent Longonot-derived Suswa pumice beds, L1 and L2 on Fig. 5, section A.

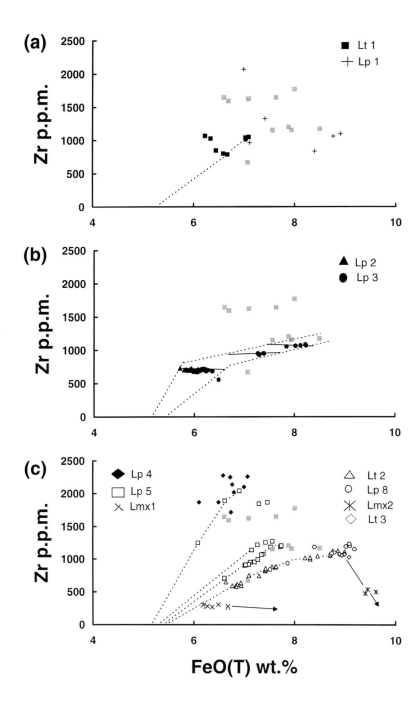

Fig. 7. Variation of Zr with total iron as FeO(T) in each stratigraphic unit of the Longonot volcanic succession. Longonot-derived Suswa samples from the Olgumi Formation (S2) are shown as stippled squares. Dashed lines show fractionation trends in zoned units. Solid lines show mixing trends.

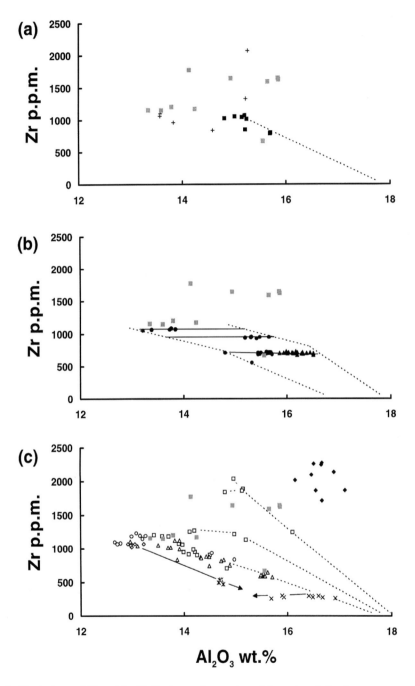

Fig. 8. Variation of Zr with Al_2O_3 in each stratigraphic unit of the Longonot volcanic succession. Longonot-derived Suswa samples from the Olgumi Formation (S2) are shown as stippled squares. All other symbols as in Fig. 7.

Table 1. Major oxide and trace-element chemistry of the Longonot-derived Suswa pumice from the Olgumi Formation.

Sample No.	IS207	IS96	IS88	IS110	IS109	IS184	IS200	IS151	IS106	IS202
Wt%										
SiO_2	59.89	60.58	58.98	59.68	59.47	61.40	61.03	60.99	61.64	59.88
Al_2O_3	15.85	13.34	14.94	15.65	15.86	15.56	14.24	13.59	13.79	14.13
TiO_2	0.36	0.47	0.47	0.38	0.35	0.65	0.50	0.46	0.48	0.44
$FeO(T)$	6.59	7.94	7.63	6.68	7.08	7.06	8.50	7.56	7.88	8.00
MnO	0.26	0.29	0.26	0.25	0.26	0.25	0.26	0.39	0.24	0.30
MgO	0.36	0.17	0.19	0.37	0.20	0.40	0.10	0.19	0.17	0.13
CaO	1.24	1.64	1.91	1.31	1.20	1.40	1.09	1.43	1.36	1.02
Na_2O	5.77	5.87	5.76	5.77	5.67	4.45	5.71	5.74	5.50	6.16
K_2O	4.90	5.29	5.08	5.40	5.32	5.27	4.97	4.83	5.18	5.06
P_2O_5	0.07	0.05	0.04	0.03	0.06	0.08	0.05	0.03	0.04	0.06
Trace element	0.34	0.25	0.33	0.33	0.33	0.16	0.25	0.24	0.26	0.36
Total	96.00	96.32	96.01	96.22	96.19	97.07	97.17	95.87	96.97	95.98
ppm										
V	22	1	1	1	1	1	1	1	1	1
Cr	1	15	6	4	9	6	10	9	19	28
Co	3	16	27	21	18	7	17	11	11	24
Ni	21	6	5	8	7	9	10	6	4	8
Cu	6									
Zn	339	288	277	330	335	195	296	273	289	341
Rb	252	185	224	237	239	132	184	184	183	246
Sr	14	8	28	15	9	11	9	11	9	15
Y	183	148	207	191	197	91	145	146	150	202
Zr	1649	1159	1647	1594	1625	671	1174	1153	1205	1774
Nb	385	273	382	388	371	171	285	276	288	418
Ba	29	14	18	21	23	48	17	19	13	23
Ce	434	336	390	382	407	214	337	296	355	414
Pb	30	26	34	29	34	15	24	23	21	36
Th	38	28	37	38	40	18	29	28	26	40

Table 2. *Ranges of major oxide and trace-element chemistry for each of the Longonot stratigraphic members*

Member	Lt1	Lp1	Lp2	Lp3	Lp4	Lp5	Lt2	Lp8	Lt3
Wt%									
SiO_2	60.23–61.95	57.18–61.95	61.04–61.93	59.43–61.99	58.79–60.13	59.48–62.63	61.27–63.00	62.26–63.49	61.43–61.97
Al_2O_3	14.82–15.71	13.57–15.27	15.71–16.51	13.21–15.72	16.15–17.11	13.42–16.10	12.96–15.71	12.67–14.98	12.92–13.22
TiO_2	0.54–0.65	0.44–0.63	0.62–0.68	0.50–0.67	0.37–0.48	0.42–0.54	0.58–0.70	0.59–0.65	0.28–0.69
FeO(T)	6.23–7.09	6.99–8.91	5.71–6.19	5.85–8.23	6.10–7.01	6.08–7.74	6.63–8.99	7.51–9.19	8.89–9.08
MnO	0.27–0.34	0.32–0.45	0.21–0.32	0.28–0.44	0.21–0.25	0.24–0.38	0.21–0.35	0.31–0.41	0.33–0.36
MgO	0.20–0.32	0.18–0.42	0.26–0.37	0.17–0.52	0.16–0.23	0.03–0.21	0.13–0.67	0.00–0.14	0.22–0.34
CaO	0.84–1.04	0.93–3.66	1.06–1.26	0.85–2.34	0.80–1.13	0.78–1.26	0.36–1.46	0.98–1.21	0.22–1.30
Na_2O	5.88–8.02	5.78–9.43	6.71–8.01	7.00–9.41	7.31–8.21	7.07–9.16	6.32–8.93	7.22–8.61	6.74–8.43
K_2O	4.76–5.64	4.58–5.26	5.10–5.18	4.54–5.72	4.60–4.99	4.59–4.94	4.38–5.68	4.42–4.85	4.62–4.76
P_2O_5	0.05–0.08	0.05–0.08	0.08–0.10	0.05–0.28	0.03–0.05	0.03–0.06	0.04–0.15	0.06–0.10	0.09–0.13
ppm									
V	18–32	20–25	25–34	20–33	12–17	14–25		24	
Co								13–20	
Ni			8–11					11–13	
Cu									
Zn	246–331	301–370	207–233	213–363	321–431	224–415	166–302	245–353	260–268
Rb	128–182	143–281	110–123	101–176	234–290	123–274	98–180	143–190	155–182
Sr	<5–9	<5–10	<5–10	11–17	<5	<5	<5–11	<5	<5–12
Y	103–138	122–230	90–97	76–161	187–237	97–225	76–148	101–156	130–139
Zr	786–1069	838–2072	674–727	554–1083	1714–2275	704–2037	575–1125	834–1231	1023–1070
Nb	204–275	240–533	179–192	140–284	436–556	173–487	130–263	198–294	242–250
Ba	<10	<10–12	<10–96	<10–185	<10	11–70	<10–254	<10	<10
Ce	204–265	240–429	162–201	138–276	384–484	196–429	171–321	250–334	291–305
Pb	14–30	23–40	15–27	14–32	37–47	20–44		21–27	
Th	16–23	21–51	11–21	12–28	38–51	17–44		19–27	

Moving to older units, Figs 7b and 8b reveal an overlap of Longonot-derived Suswa pumice containing Zr < 1200 ppm, and the Longonot Lp3 trend. With the exception of Na_2O, all other oxides and elements show similar ranges for the two suites (Tables 1 & 2). Oxide totals as low as 95.87 wt% for the Longonot-derived Suswa samples suggest that disparate Na_2O ranges may result from a possible loss of Na_2O during hydration of the Suswa samples (Noble 1967; Macdonald & Bailey 1973). This group of samples was collected from the two youngest Longonot-derived Suswa pumice beds, L3 and L4 (Fig. 5), whilst the group of samples with Zr > 1400 ppm were collected from pumice beds L1 and L2. The high Zr group must therefore be derived from Longonot units erupted earlier than the low Zr group.

Variation within the oldest exposed Longonot units are shown on Figs 7a and 8a. These plots record a general overlap of the Longonot-derived Suswa pumice and the Longonot Lp1 spread. Once again, with the exception of Na_2O, all other oxides and elements show similar ranges for the two suites.

The relationships displayed on Figs 7 and 8 suggest two possible correlations. Either all the Longonot-derived Suswa pumice beds are equivalent to the Longonot Lp1 member, or alternatively, the early, high Zr Longonot-derived Suswa pumice beds L1 and L2 are equivalent to the Longonot Lp1 member whilst the later, lower Zr Longonot-derived Suswa pumice beds L3 and L4 are equivalent to the Longonot Lp3 member. In both cases, the correlation indicates synchronized initiation of incremental caldera collapse at Longonot and Suswa, accompanied by the emplacement of ignimbrites. This suggests that caldera formation is a regional rather than local event in the southern part of the rift floor, highlighting the potential role for regional tectonics as a controlling mechanism.

The role of regional tectonics

Longonot and Suswa both display alignments of their eruptive vents and fissures (Fig. 9). At Suswa, a N–S orientated major alignment is defined by the Tandamara vent, the Enkorika Fissure Zone, the centres of calderas 1 and 2 (A on Fig. 9a) and the pit crater on the Ol Doinyo Onyoke cone (B on Fig. 9a). The alignment is parallel to the regional N–S normal faults exposed on the rift floor south of Suswa (Baker et al. 1988), and can be traced northwards into one of the many N–S fractures cutting the Olkaria trachyte – comendite centre. At Longonot, a similar NW–SE orientated major alignment is defined by vents of the Lower Trachyte Member (Lt2) of the Longonot Trachyte Formation, which includes the primary vent at the summit of the lava pile (A on Fig. 9b). This alignment is parallel to the regional NW–SE normal faults of the rift margin and floor to the E and NE of Longonot (Clarke et al. 1990). Minor alignments of eruptive vents at Longonot are centred on the lava pile vent (A on Fig. 9b) or the later pit crater (B on Fig. 9b), and are related to the local stress field developed above the shallow magma chamber.

In typical fissure swarms of SW Iceland, Forslund & Gudmundsson (1991) recognized that surface dilation is almost entirely due to tensional fractures and vertical normal faults whilst at depths greater than 1 km the contribution of dykes in accommodating extension becomes more dominant. This highlights the important role of magma influx as an accommodation mechanism during periods of extension. Gudmundsson (1995) emphasized that in any particular fissure swarm, the number of tension fractures exceeds that of normal faults, again reflecting the dominant

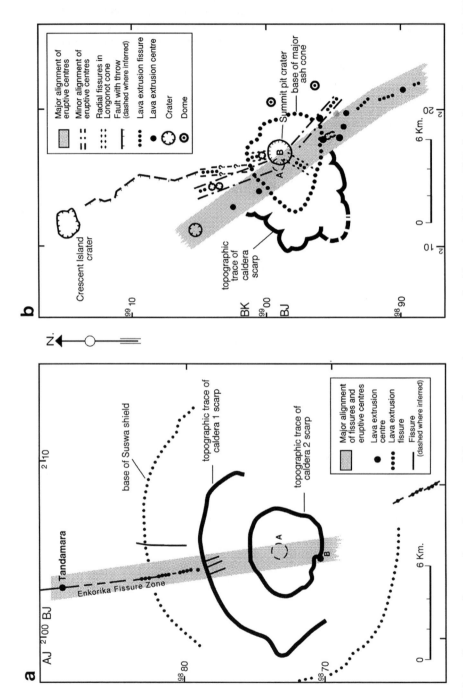

Fig. 9. (a) Major alignment of exposed eruptive vents and fissures at Suswa. A = centre of calderas 1 and 2; B = Ol Doinyo Onyoke cone; grid = UTM Zone 37M. (b) Major alignment of exposed Lt2 eruptive vents and fissures at Longonot. A = summit of Lt2 lava pile; B = centre of pit crater; grid = UTM Zone 37M.

magmatic accommodation of extension at depth. Superimposed on the regional extensional stress field are local stress fields developed above shallow level crustal magma chambers. These give rise to concentric, inwardly dipping dyke sheets (cone sheets) up to 9 km in radius and may have facilitated caldera formation (Gautneb & Gudmundsson 1992). A similar situation occurs in the north Kenya Rift where extensive basalts cropping out between silicic centres were erupted from tensional fissures and from fault-fissures with vertical fault displacements (Weaver 1978). Faulting and basalt extrusion appears to have been contemporaneous, suggesting some accommodation of the extension by buoyant rise and injection of basalt from the crust–mantle boundary. The lack of extensive fissure basalts between silicic centres in the south Kenya rift may reflect a greater crustal thickness (KRISP WORKING PARTY 1991; Keller *et al.* 1994; Mechie *et al.* 1994; Morley 1994) and a lower rate of extension (Hendrie *et al.* 1994; Latin *et al.* 1993) compared to the north rift. Within the upper crust of the south Kenya Rift, accommodation of extension by magma influx may be facilitated by draining of basalt and trachyte from high level crustal reservoirs below central volcanoes rather than by buoyant rise of basalt from the crust–mantle boundary.

Parallelism of major volcano alignments and regional faults at Longonot and Suswa suggests that regional tension fractures or faults which intersect the shallow-level magma chambers, are utilized as conduits for transporting magma from the chambers. Opening of tension fractures or movement on normal faults, in response to an episode of regional extension, may have provided the trigger for synchronous lateral magma withdrawal along regional fractures, decompression of the magma chambers and caldera collapse at Longonot and Suswa. Support for such a mechanism is furnished by the calculated caldera volume and syn-caldera erupted magma volume at each centre. For Suswa, Skilling (1993) recorded a caldera volume of $22 \, km^3$ and a syn-caldera erupted magma volume of $5.2 \, km^3$, whilst for Longonot, Scott (1980) calculated a caldera volume of $26.5 \, km^3$ and a syn-caldera erupted magma volume of $11.25 \, km^3$. In both cases the caldera volume is significantly greater than the volume of magma erupted during syn-caldera activity, a volume relationship consistent with lateral withdrawal of magma along regional fractures during caldera formation.

At the basalt–trachyte volcano Silali in the northern sector of the Kenya Rift, Smith *et al.* (1995) identified links between regional extension and volcanic activity. They argued that regional extension was probably the cause of fracturing and decompression of the magma chamber around 133–131 ka, resulting in a temporary cessation of activity until *c*. 120 ka. They also suggested that caldera formation can be linked to lateral drainage of basaltic magma from a high level central reservoir, a mechanism similar to that proposed by Sparks *et al.* (1981) for the Askja volcano, Iceland. For Emuruangogolak volcano north of Silali, Weaver (1978) draws attention to fault scarps cutting the early trachyte shield which are in a similar state of degradation to the caldera 2 scarp, suggesting that caldera collapse may have again been coeval with a period of normal faulting. Additional evidence for a possible link between regional tectonics and caldera collapse is provided by the calderas at Menengai (Leat 1984) and Kilombe (McCall 1964). Sector graben showing two major orientation directions, each parallel to a regional rift fault orientation, form an integral part of the caldera scarp at each centre, the sector graben at Kilombe showing alignments which intersect near the centre of the

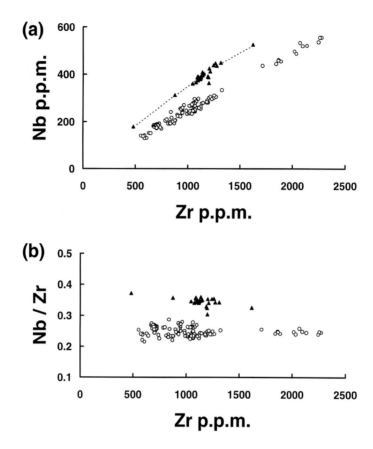

Fig. 10. (**a**) Variation of Nb with Zr for all the analysed volcanics from Longonot and the Suswa Olgumi Formation (S2). Dashed line shows the dominant trend of the Suswa S2 volcanics, which define a coherent fractionation trend slightly curved towards the right of the graph, reflecting the gradually decreasing Nb/Zr ratio shown in (b). (**b**) Variation of Nb/Zr with Zr for all the analysed volcanics from Longonot and the Suswa Olgumi Formation (S2). Filled triangles – Suswa S2 volcanics; open circles – Longonot volcanics.

caldera. These features suggest a possible role for fault intersection in the control of magma reservoir location.

Another possible effect of regional extension episodes is illustrated in Figs 10a and 10b which show plots of Nb against Zr and Nb/Zr against Zr for all the analysed Longonot volcanics, and for the early syn-caldera Olgumi Formation (S2) trachyte pyroclastics from Suswa. Apart from three samples, the Suswa S2 sample array defines a tightly constrained, linear trend on both plots. The three samples which lie off this trend define a subsidiary trend towards a lower Nb/Zr ratio. A small amount of mixing with Longonot-derived trachyte magma may have occurred by lateral propagation of magma along an active tension fracture or normal fault during the initial stages of synchronous caldera collapse. As noted above, however, the major alignment of eruptive vents on Suswa extends northwards through the Tandamara vent into one of the many N–S fractures cutting the Olkaria complex. During the initial stages of caldera formation at Longonot and Suswa, lavas and pyroclastics of the Lower Comendite member (O2) were erupted at Olkaria (Clarke *et al.* 1990). The volcanics of this stratigraphic unit define a coherent linear trend on a plot of Nb against Zr, and display a Nb/Zr ratio lower than that of the Longonot volcanics (Clarke *et al.* 1990). Lateral propagation of Olkaria O2 comendite magma southwards towards Suswa along a N–S tension fracture, and minor mixing with Suswa trachyte magma during or prior to initiation of caldera collapse may provide an explanation for the Suswa S2 subsidiary trend, which is more consistent with the available geological and trace-element evidence.

Conclusions

1. The eruptive products of the Kenyan Quaternary volcanoes Longonot and Suswa can be uniquely characterized by their contrasting Nb/Zr ratios. This has led to the identification of Longonot-derived airfall pumice beds within the Suswa early syn-caldera Olgumi Formation (S2).
2. Tephrachronology using geochemical fingerprinting of the Longonot-derived Suswa pumice beds demonstrates synchronous initiation of caldera formation at the two centres. This suggests that caldera formation is a regional rather than local event in this southern part of the rift, highlighting the potential role of regional tectonics as a controlling mechanism.
3. Major vent alignments at each centre are parallel to regional fault trends whilst the volume of material consumed during caldera formation is significantly greater than magma volumes erupted during syn-caldera activity. These features suggest that an episode of regional extension in the southern rift may have been the controlling mechanism triggering synchronous lateral magma withdrawal along regional fractures, decompression of the magma chambers and caldera collapse.
4. At Suswa, a small amount of mixing of S2 trachyte magma with O2 comendite magma from the Olkaria centre to the north, occurred during or prior to the initiation of caldera collapse. Mixing was facilitated by the lateral propagation of comendite magma southwards along a N–S tension fracture (the Tandamara line) which links the two centres.

We are grateful to R. Macdonald, D.K. Bailey and L.A.J. Williams for their help and constant encouragement during our Kenyan work. Initial work at Longonot was undertaken by SCS during the tenure of a NERC Research Studentship. Fieldwork undertaken by SCS in 1986 was supported by The Royal Society of London. Fieldwork by SCS in 1987 was supported by The Nuffield Foundation and undertaken in collaboration with field staff involved in a programme of exploration for geothermal energy in southern Kenya, funded by the UK Overseas Development Administration. Work on Suswa was undertaken by IPS during tenure of an NERC Research Studentship.

References

BAKER, B. H. & HENAGE, L .F. 1977. Compositional changes during crystallization of some peralkaline silicic lavas of the Kenya Rift Valley. *Journal of Volcanology and Geothermal Research*, **2**, 17–28.
——, MITCHELL, J. G. & WILLIAMS, L. A. J. 1988. Stratigraphy, geochronology and volcano-tectonic evolution of the Kedong–Naivasha–Kinangop region, Gregory Rift Valley, Kenya. *Journal of the Geological Society, London*, **145**, 107–116.

CAMBRAY, F. W., VOGEL, T. A. & MILLS, J .G. 1995. Origin of compositional heterogenieties in tuffs of the Timber Mountain group: the relationship between magma batches, and magma transfer and emplacement in an extensional environment. *Geophysical Abstracts in Press*, **5**(3).

CLARKE, M. C. G., WOODHALL, D. G., ALLEN, D. & DARLING, G. 1990. Geological, volcanological and hydrogeological controls on the occurrence of geothermal activity surrounding Lake Naivasha, Kenya. British Geological Survey – Kenya Ministry of Energy report, Nairobi.

FORSLUND, T. & GUDMUNDSSON, A. 1991. Crustal spreading due to dykes and faults in southwest Iceland. *Journal of Structural Geology*, **13**, 443–457.

GUDMUNDSSON, A. 1995. Infrastructure and mechanics of volcanic systems in Iceland. *Journal of Volcanology and Geothermal Research*, **64**, 1–22.

GAUTNEB, H. & GUDMUNDSSON, A. 1992. Effect of local and regional stress fields on sheet emplacement in west Iceland. *Journal of Volcanology and Geothermal Research*, **51**, 339–356.

HENDRIE, D. B., KUSZNIR, N. J., MORLEY, C. K. & EBINGER, C. J. 1994. Cenozoic extension in northern Kenya: a quantitative model of rift basin development in the Turkana region. *Tectonophysics*, **236**, 409–438.

JOHNSON, R. W. 1969. Volcanic geology of Mount Suswa, Kenya. *Philosophical Transactions of the Royal Society of London*, **A265**, 383–412.

JONES, W. B. 1981. Chemical effects of deuteric alteration in some Kenyan trachyte lavas. *Mineralogical Magazine*, **44**, 279–285.

KELLER, G. R., PRODEHL, C., MECHIE, J., FUCHS, K., KHAN, M. A. ET AL. 1994. The East African rift system in the light of KRISP 90. *Tectonophysics*, **236**, 465–483.

KRISP WORKING PARTY 1991.Large-scale variation in lithospheric structure along and across the Kenya rift. *Nature*, **354**, 223–227.

LATIN, D., NORRY, M. J. & TARZEY, R. J. E. 1993. Magmatism in the Gregory Rift, East Africa: evidence for melt generation by a plume. *Journal of Petrology*, **34**, 1007–1027.

LEAT, P. T. 1984. Geological evolution of the trachytic caldera volcano Menengai, Kenya Rift Valley. *Journal of the Geological Society*, **141**, 1057–1069.

MACDONALD, R. & BAILEY, D. K. 1973. The chemistry of the peralkaline oversaturated obsidians. *U. S. Geological Survey Professional Paper*, **440-N-1**, 1–37.

MAHOOD, G. A. & STIMAC, J. A. 1990. Trace-element partitioning in pantellerites and trachytes. *Geochimica et Cosmochimica Acta*, **54**, 2257–2276.

MCCALL, G. J. H. 1964. Kilombe caldera, Kenya. *Proceedings of the Geologists Association*, **75**, 563–572.

MECHIE, J., KELLER, G. R., PRODEHL, C., GACIRI, S., BRAILE, L. W. ET AL. 1994. Crustal structure beneath the Kenya Rift from axial profile data. *Tectonophysics*, **236**, 179–200.

MORLEY, C. K. 1994. Interaction of deep and shallow processes in the evolution of the Kenya rift. *Tectonophysics*, **236**, 81–91.

NOBLE, D. C. 1967. Sodium, potassium, and ferrous iron contents of some secondarily hydrated natural silicic glasses. *American Mineralogist*, **52**, 280–286.

SCOTT, S. C. 1980. The geology of Longonot volcano, Central Kenya: a question of volumes. *Philosophical Transactions of the Royal Society of London*, **A296**, 437–465.

SKILLING, I. P. 1993. Incremental caldera collapse of Suswa volcano, Gregory Rift Valley, Kenya. *Journal of the Geological Society, London*, **150**, 885–896.

SMITH, M., DUNKLEY, P. N., DEINO, A., WILLIAMS, L. A. J. & MCCALL, G. J. H. 1995. Geochronology, stratigraphy and structural evolution of Silali volcano, Gregory Rift, Kenya. *Journal of the Geological Society, London*, **152**, 297–310.

SPARKS, R. S. J., WILSON, L. & SIGURDSSON, H. 1981. The pyroclastic deposits of the 1875 eruption of Askja, Iceland. *Philosophical Transactions of the Royal Society of London*, **A229**, 241–273.

WEAVER, S. D. 1978. The Quaternary caldera volcano Emuruangogolak, Kenya Rift, and the petrology of a bimodal ferrobasalt–pantelleritic trachyte association. *Bulletin Volcanologique*, **40**, 209–230.

——, GIBSON, I. L., HOUGHTON, B. F. & WILSON, C. J. N. 1990. Mobility of rare earth and other elements during crystallization of peralkaline silicic lavas. *Journal of Volcanology and Geothermal Research,* **43**, 57–70.

WÖRNER, G., BEUSEN, J.-M., DUCHATEAU, N., GIJGELS, R. & SCHMINCKE, H.-U. 1983. Trace element abundances and mineral/melt distribution coefficients in phonolites from the Laacher See Volcano (Germany). *Contributions to Mineralogy and Petrology,* **84**, 152–173.

Deep sea tephra from Nisyros Island, eastern Aegean Sea, Greece

J. C. HARDIMAN

Department of Earth Sciences, Downing Street, Cambridge CB2 3EQ, UK (Present address: British Geological Survey, Keyworth, Nottingham NG12 5GG, UK)

Abstract: Isopach maps have been constructed onland for the proximal deposits of the caldera-forming pyroclastic eruptions from the Quaternary volcano Nisyros and are augmented by a newly recognized Nisyros ash layer in the SE Aegean. The Lower Pumice eruption had a NE–SW dispersal direction with onland isopachs from 1–7 m thickness and a vent position on the eastern side of the ancestral cone. A 3 mm thick layer of Lower Pumice ash is found 75 km away to the SW of the island at a depth of 133 cm. This layer yields a Lower Pumice age estimate of 35 ka. The Upper Pumice had a NNE dispersal direction with onland isopachs from 1–7 m thickness and a vent position on the northern side of the island. No Upper Pumice ash has been found in the core record. Preliminary volume estimates derived from the isopach maps and the lack of regionally correlatable Nisyros ash deposits in the Aegean core record suggest that the caldera formation on the island was incremental and tephrostratigraphic markers suggest it occurred between *c.* 35 and 31 ka ago.

Nisyros lies at the eastern end of the Aegean Volcanic arc and is one of at least 8 centres active during the Quaternary (Fig. 1). The Nisyros centre (Fig. 2) is a small (42 km^2) symmetric composite stratovolcano built up from the eruption of basaltic–andesite to rhyolite lavas and pyroclastics (Di Paola 1974). Destruction of this stratocone occurred as a result of three large eruptions of rhyolite composition: the Lower Pumice and Upper Pumice pyroclastic members and a lava flow, the Nikia Rhyolite which caused the formation of a large (10 km^2) central caldera (Di Paola 1974). Previous isopach and isopleth maps of the proximal units of these two pyroclastic eruptions (Limburg & Varekamp 1991) and tephrostratigraphic works (e.g. Vinci 1983, 1985) have suggested that the Lower and Upper Pumice eruptions were of a magnitude (>1 km^3 eruptive volume) such that the ash deposits should appear in the Aegean tephrostratigraphic record. After the Upper Pumice eruption, a suite of NE–SW trending rhyodacite domes have been extruded onto the western side of the caldera floor and the island currently shows hydrothermal activity (Marini *et al.* 1993).

The volcano has formed entirely within the Quaternary with the oldest known date for the stratocone build up given by the occurrence of the *c.* 145 ka Kos Plateau Tuff overlying the submarine base of the volcano (Rehren 1988). Ages of 38 ka and 66 ka have been obtained on dacite and rhyolite lavas, respectively, occurring

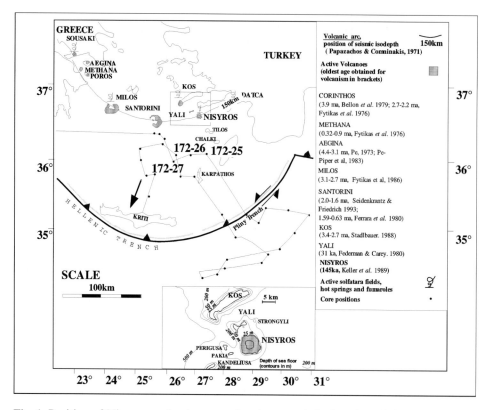

Fig. 1. Position of Nisyros on the Aegean volcanic arc and the location of the cores used in this study on the route of the *R/V Trident* 172 Cruise, 17 Sept.–3 Oct., 1975 (Rhode Island). Inset map shows the positions of the sub-group of volcanic islands surrounding Nisyros and the bathymetric contours around the island in metres.

midway up the stratocone (Rehren 1988; Keller *et al.* 1989). The ages obtained for the destructive activity have so far been contradictory. An age of 24 ka has been obtained from charcoal in a palaeosol beneath the Lower Pumice (^{14}C uncorrected age, Rehren 1988), and an age of 31 ka has been obtained for the Yali-2 layer (^{18}O stratigraphy, Federman & Carey 1980), which stratigraphically overlies the Upper Pumice on the island and whose deposits are found inside the caldera abutting against the base of the northern caldera wall (Rehren 1988).

The volcanic history and stratigraphy of the island have been described in works by Di Paola (1974); Rehren (1988); Keller *et al.* (1990); Limburg & Varekamp (1991); Vougiokalakis (1992) and Francalanci *et al.* (1995). Di Paola (1974) produced the first large scale (1:25 000) geological map of Nisyros and first outlined the stratigraphy and volcanic evolution of the island that has been used as the basis for more recent work. A complete set of whole rock and glass analyses of the Nisyros and Yali stratigraphy is contained within Rehren (1988). Keller *et al.* (1990) and Limburg & Varekamp (1991) have both summarized the silicic pyroclastic stratigraphy of the island. Keller *et al.* (1990) included a detailed description of the Panaghia Kyra stratocone silicic pyroclastic sequence and Limburg & Varekamp (1991) gave a volcanological interpretation for the Upper Pumice eruption.

Vougiokalakis (1992) produced descriptions of the entire pyroclastic and lava stratigraphy of the island with a new geological map of Nisyros at 1:5000 scale. Francalanci et al. (1995) supplemented the knowledge gained from the previous works on the stratigraphy and geochemistry of Nisyros with trace element and isotopic data leading to a combined petrogenetic model for the evolution of the island.

Here, a summary of the stratigraphy of the Lower and Upper Pumice eruptions and the postulated isopach maps for the tephra fall phases are presented. The search for Nisyros ash in cores from the surrounding Aegean seafloor leads to the first finding of Nisyros ash in the tephrostratigraphic record (excepting, Vinci 1985). The results are used to confirm the possible areal distribution of the caldera phase silicic eruptions and provide a distal extrapolation.

The Aegean cores used to look for Nisyros ash were collected during the *R/V Trident* 171 and 172 expeditions (1975) and are currently kept at the University of Rhode Island (Fig. 1, Watkins *et al.* 1978; Thunell *et al.* 1979; Federman & Carey 1980). Correlations with onland tephra sources and the known tephrochronological record in the vicinity were established using criteria of Munsell colour and mineralogy prior to electron microprobe analysis of glasses using EDS Spectrometry. EDS analysis requires analytical conditions to be optimized in order to reduce alkali loss to a minimum (Lineweaver 1962; Neilsen & Sigurdsson 1981). The rhyolite tephras of Kos and Nisyros are most distinguishable by their Na_2O and K_2O values so it was considered important to get accurate analyses for these elements. The glass analyses of the calc-alkaline andesite to rhyolite tephra of the Aegean sources are also fairly similar and since comparisons between the compositions of hydrated glass are made after analyses have been recalculated to 100% (anhydrous), each element was determined as accurately as possible so that artefacts (due to volatile loss, for example) were not introduced during recalculation. The tephras found are placed in the context of the previously established regionally correlated tephra markers in the Aegean sea: the Santorini Minoan (Z2), Santorini Cape Riva (Y2), Italian Peninsula Campanian (Y5), and Kos Plateau Tuff (W3; Keller 1971, 1980, 1981; Thunell *et al.* 1977; Watkins *et al.* 1978; Keller *et al.* 1978; Thunell *et al.* 1979; Federman & Carey 1980; Vinci 1985; using the terminology of Keller *et al.* (1978)).

Summary of the pyroclastic stratigraphy of Nisyros

The stratigraphy of the Lower and Upper Pumice eruptions is summarized below and their exposure regions are outlined in the geological map in Fig. 2. The earlier sequence of silicic pyroclastics which occurred during the stratocone build-up, the Panaghia Kyra Series, has been laterally correlated with its tephra layers found on the neighbouring islands of Chalki, Tilos and the Datia peninsula, Turkey just a few kilometres distant (Keller et al. 1990). However the deduced easterly dispersal of these eruptions suggests there is little possibility of their appearance in the Trident cores which were restricted to a region south of Nisyros (Fig. 1; Keller et al. 1990). No deposits from the Lower and Upper Pumice eruptions have been found on neighbouring islands.

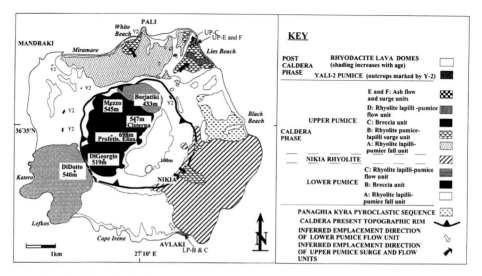

Fig. 2. Simplified geological map of Nisyros showing the distribution of the silicic pyroclastic deposits and the eruptions since the start of the caldera formation on the island (unshaded = stratocone lavas and pyroclastics).

Lower Pumice (LP)

The Lower Pumice deposits unconformably overlie the highest stratigraphic units on the island flanks which are dacite lavas in the northern part of the island and a sequence of andesitic pyroclastics elsewhere (Vougiokalakis 1992). No exposures occur in the caldera wall and the Lower Pumice eruption is considered to have started the caldera formation on the island. The LP member subdivides into three units with different inferred depositional mechanisms: lapilli-pumice fall (LP-A), breccia (LP-B) and lapilli-pumice flow (LP-C).

- **LP-A** deposits cover the stratocone slopes from the caldera rim to the coast on all sides of the island but outcrops are laterally separated by the later Nikia Rhyolite flows on the southeast flank and by the post-caldera lava dome 6 (DiDotto) on the southwest flank (Fig. 2). These separate outcrops have been previously named 'Pumice Unit 4' in the north and the Stavros Pumice or 'Pumice Unit 3' in the south (Limburg & Varekamp 1991).
- **LP-A1** is a millet-seed lithic sub-unit which forms a thin basal layer to LP-A over the whole island. It is approximately 2 cm thick in the southern part of the island and 1–2 mm thick in the north. LP-A1 is comprised of lithic clasts up to 2 mm diameter of the same composition as those in the overlying fall units and is considered to represents lithics from the vent opening phase deposited by fallout.
- **LP-A2** is an ash sub-unit which conformably overlies LP-A1. It is approximately 5 cm thick in the south of the island and 2–4 cm thick in the north and east. The ash is pale-grey, normally oxidized to pink. LP-A2 represents the deposits from the first phase of magmatic activity and is also assumed to have been deposited by fallout.
- **LP-A3** is an ungraded white lapilli-pumice fall deposit. The thickness of LP-A3

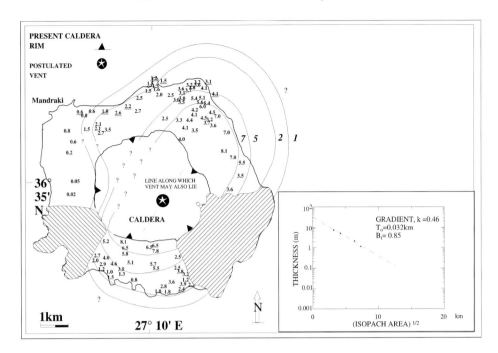

Fig. 3. Lower Pumice (LP-A) isopach map. Thicknesses (m) are underlined where the base and top of the unit are exposed, the others given are minimum, and superimposed are the postulated isopach contours and possible vent position assuming that the vent lies at the projected maximum thickness. Inset shows a plot of (ln) thickness against isopach area ($^{1/2}$). T_o is the maximum thickness of the proximal part extrapolated to source, k is the gradient and B_t is the thickness decay constant, defined by Pyle (1989).

is given in Fig. 3. Changes in the characteristics of the unit are seen around the island flanks. In the north and in the localities along the south coast the deposit is clast supported and matrix poor (c. 1 vol.%). Lithics are < 10 cm across and only up to 10 modal % abundance. On the southern flanks close to the caldera the deposit is slightly indurated with clasts more poorly sorted and matrix supported in fine ash (c. 5 vol.%). Lithics are up to 80 cm across and up to 50 modal % abundance. In the north, LP-A3 is weathered to light orange or cream where it is overlain by any of the Upper Pumice units, whilst in the south and east the deposit is unweathered (whitish). In the northwest the deposits are reworked with well rounded pumice clasts cemented by ash and crystals and the lithics are concentrated in 10–20 cm thick layers. Thickness of LP-A3 decreases southwards on the western slopes until pumice clasts are seen only in the surface layer of the soil. The deposits are also often eroded inland particularly in the northeast where the Lower Pumice often forms the highest stratigraphic unit. The pumice contains phenocrysts of plagioclase, alkali feldspar, quartz, hypersthene, magnetite, ilmenite, apatite and zircon in a glassy matrix.

- **LP-B** is a c. 1.5 m thick layer of lithic breccia exposed between LP-A and LP-C at the coast between Avlaki and the Nikia Rhyolite. It consists of unsorted randomly orientated, angular clasts with a modal clast diameter of 20 cm within a brown ash matrix. Clasts include lavas with compositions corresponding to

the southern flank of ancestral cone, sedimentary basement clasts, skarns and igneous intrusives, up to 70 cm across, and well rounded pumices up to 40 cm across. The exposure of LP-B is extremely limited onland which makes it difficult to determine if it represents a lag breccia or a ground breccia.
- **LP-C** is a lapilli-pumice unit with a minimum onland thickness of 4 m exposed in small cliffs at the coast between Avlaki and the Nikia Rhyolite. Thickness increases to 10 m in a cliff 50 m further east at the border with the Nikia Rhyolite but most of this exposure is reworked and only the bottom 2 m is primary. Further deposits are presumed to continue eastwards underlying the brecciated base of the Nikia Rhyolite flows. LP-C is a matrix-supported deposit with rounded pumice and lapilli clasts loosely consolidated by ash and crystals. It has a typical pumiceous ignimbrite appearance and is considered to have been deposited by a flow mechanism. The lowest 20 cm of the unit consists of indurated fine-grained ash with flattened lapilli with an undulose but sharp base overlying LP-B. The plane-parallel lamination and indurated ash in this part of the deposit appear to indicate deposition by a surge mechanism, possibly due to an episode of phreatomagmatic activity. In this case, therefore, the underlying unit LP-B would have to represent the lag breccia from a different flow. The other interpretation is that the base represents the deposits from a different portion of the LP-C flow, for instance either the head which deposited first or 'erosive' fine-grained deposits from the flow base. In all cases the flattening of the lapilli in this layer is considered to be due to deposition of the overlying pumice-lapilli portion before the deposit had cooled. Thicknesses and the limited exposure suggest that LP-C was directed southeastwards.

The Lower Pumice is overlain by the Nikia Rhyolite on the southeastern slopes, by a 10–40 cm thick palaeosol beneath the Upper Pumice fall unit on the northern slopes and an up to 30 cm thick palaeosol beneath the Upper Pumice surge unit at Lies Beach on the east coast. The ashy soil changes colour from light brown to orange and is oxidized to purple for a few centimetres at the top where it was heated by the overdeposition of the Upper Pumice.

Upper Pumice (UP)

UP deposits mantle the Lower Pumice on the northern and eastern flanks and the Nikia Rhyolite in the southeast of the island. The UP member consists of the following units with inferred depositional mechanisms: lapilli-pumice fall (UP-A), pumiceous surge (UP-B), lithic-lag breccia (UP-C), pumiceous flow (UP-D) and phreatomagmatic units (UP-E and UP-F; Limburg & Varekamp 1991). The Upper Pumice has been previously named the 'Upper Caldera Pumice', (Keller et al. 1990) and 'Unit 5', (Limburg & Varekamp 1991).

- **UP-A1** is a well-sorted, 4–10 mm thick, layer of millet seed lithics, presumed to have been emplaced by fallout and to represent deposits from the opening of the vent.
- **UP-A2** is a 4 cm thick white ash layer that is oxidized to purple over the top centimetre, also presumed to have a fallout origin and to represent the first deposits from the magmatic phase of the eruption. UP-A1 and UP-A2 are

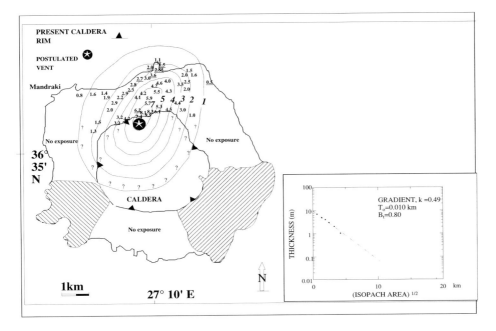

Fig. 4. Upper Pumice (UP-A) isopach map. Thicknesses (m) are underlined where the base and top of the unit are exposed, the others given are minimum, and superimposed are the postulated isopach contours and possible vent position assuming that the vent lies at the projected maximum thickness. Inset shows a plot of (ln) thickness against isopach area $(^{1/2})$. T_o is the maximum thickness of the proximal part extrapolated to source, k is the gradient and B_t is the thickness decay constant, defined by Pyle (1989).

always exposed conformably together, and are exposed everywhere that the base of the Upper Pumice is exposed.

- **UP-A3** is an ungraded, unconsolidated, clast-supported, rhyolitic lapilli-pumice unit, with a matrix of 4 vol.% crystals and ash. The thickness of UP-A3 is given in Fig. 4. Pumice clasts are subrounded with the sub-angularity of some clasts reflecting break-up on deposition. Clasts are invariably fresh and white, with the larger pumices having pink oxidized interiors. The whole deposit is lithic poor (<1 modal % abundance) although approximately the basal 20 cm has lithics up to 10 modal % abundance. The pumice is crystal rich with phenocrysts of plagioclase, alkali feldspar, hypersthene, diopside, hornblende, ilmenite, magnetite, apatite and zircon in glass.
- **UP-B** is a lapilli-pumice unit that overlies UP-A on the northern side of the island from Mandraki village eastwards to Lies Beach on the east coast, laterally replacing it eastwards. It is also found in isolated places above the Nikia Rhyolite further south. The best exposures occur in the cliff along Lies Beach and in the cliff above White Beach on the west side of the Pali promontory. Onland thicknesses range up to 8 m with greatest values in lobes towards the NW and ENE in orientations either side of the fall unit maximum isopachs. Upper Pumice-B has a lower and upper facies, labelled UP-BA and UP-BB respectively.
- **UP-BA** consists of plane-parallel layers of sub- to well-rounded, poorly

vesicular lapilli-pumice clasts suspended in a well consolidated matrix of ash and coarse crystals, up to 40 cm thick, that alternate with layers of coarse ash and crystals, up to 15 cm thick. The pumice clasts can be up to 20 cm in diameter but are usually of maximum size 10 cm. The layers are laterally continuous over tens of metres with lensoid and discontinuous intercalations. The unit is lithic poor but where lithics are present they are coarser grained than in the underlying fall unit, and they are slightly more concentrated towards the base of the unit. Several large thicknesses of UP-BA occur on the flanks southeast and southwest of Pali.

- **UP-BB** consists of undulose layers, defined by thin (<5 cm) bands of lapilli intercalated with fine indurated ash and crystals. The bedding shows structures including cross lamination, climbing ripples and dunes. This upper facies is only exposed at Lies beach and beneath Pali promontory where the most complete successions of UP-B are shown.

The well rounded and low vesicularity clasts, reverse grading, and high percentage of matrix in the deposit and the coarse grain size of the lithics suggest that the UP-B facies were emplaced by a flow mechanism as dilute pumiceous surges.

- **UP-C** is a breccia unit which outcrops in one c. 1 m thick outcrop in the road cutting around the headland west of Lies Beach. It consists of angular to sub-angular lithics of stratocone lavas and pyroclastics up to 60 cm across, basement clasts up to 10 cm across, magmatic inclusions and well rounded pumices, up to 20 cm across, matrix supported in ash. UP-C has been interpreted by Limburg & Varekamp (1991) to represent a lithic-lag breccia associated with UP-D. In similarity to the Lower Pumice the breccia exposure is small onland and might suggest that it is of low volume overall.

- **UP-D** is a structureless lapilli-pumice unit exposed on the northeast flanks of the island. A thickness of 5.1 m occurs at Lies Beach on the east coast. It consists of sparse (5% modal abundance) well rounded poorly-vesicular pumice clasts, supported in a matrix of loosely consolidated lapilli, ash and crystals. Occasionally faint parallel bedding and some induration is shown. The well rounded nature of the clasts and high proportion of matrix in the deposit with a characteristic appearance of ignimbrite suggest a flow mechanism and the limited exposures indicate that the depositional direction was ENE.

- **UP-E** is a grey lapilli-ash and crystal unit of minimum thickness 6 m exposed on the northeast flanks around Lies Beach. UP-E irregularly slumps or penetrates as veins into UP-D. The overall fine grain size and absence of pumice, preservation of the original magma colour and lack of vesicularity in the clasts suggests a different fragmentation mechanism to the preceding units and UP-E is considered to have been produced by a phreatomagmatic mechanism. The limited exposures suggest that UP-E also had an ENE depositional direction.

- **UP-F** is a white fine-grained ash unit. It also contains 1 mm size accretionary lapilli (light grey ash with pink interiors), concentrated in 5–20 cm thick layers, that are often preferentially weathered out. UP-F is also interpreted to have resulted from phreatomagmatic activity with a water:magma ratio that increased further after UP-E.

The Upper Pumice on the north coast on the promontory near Pali is overlain after a 60 cm layer of palaeosol by a thin veneer of Yali 2 pumice (Rehren 1988). Where the Yali-2 pumice layer is absent, a palaeosol of up to 2 m thickness may be developed above the Upper Pumice.

Fall unit isopach data

Isopach contours for the Lower Pumice fall unit (LP-A3) may be traced onland from 7 to 1 m (Fig. 3). The contours have been extrapolated across the present caldera. Using the observation that many tephra fall deposits display exponentially decreasing thickness away from the vent (Thorarinsson 1954), the data are plotted on a (ln) thickness against isopach area (1/2) graph (Pyle 1989; inset, Fig. 3). Minimum volume estimates may be gained by the integration of the volume under the straight line graph obtained if decay is exponential (Pyle 1989; Fierstein & Nathenson 1992). The Lower Pumice onland isopach data appear also to fall on a straight line, indicating exponential decay is obeyed with a thickness decay constant (the distance over which the thickness halves) of $B_t = 0.85$ km and a thickness extrapolated to the vent, $T_o = 0.032$ km (inset, Fig. 3). This leads to an estimate of the minimum tephra fall volume of 0.30 km^3 (inset, Fig. 3). To account for the effect of the extrapolation of the onland isopach data offshore where thickness decay might not continue to be exponential, the proportion of the volume calculated to lie within the mapped isopachs is approximately 80% (Pyle 1995). Therefore the minimum deposited volume represented by the proximal material onland is 0.24 km^3 which is 80% of the proximal decay line. Since it is well known that fall deposits show an inflexion point (Pyle 1989; Fierstein & Nathenson 1992), the question of the total volume is dependent on how the isopachs extrapolate distally.

Isopach contours for the Upper Pumice fall unit can also be constructed onland from 7 to 1 m (Fig. 4). The contours are extrapolated into the caldera with approximately the same ellipticity as shown in the exposures on the flanks. The data on the coast and on the northeast part of the island are true thicknesses where the deposit is underlain by the LP-A and overlain by the UP-B. The thicknesses in the west, however, have no upper limit and must be regarded as minimum, although the values are comparable to the others. In similarity to the Lower Pumice if the isopach data are plotted as (ln) thickness versus isopach area $^{(1/2)}$ the data fall on a straight line, indicating exponential thickness decay is obeyed with $B_t = 0.8$ km, $T_o = 0.010$ km and the minimum tephra volume calculated as 0.08 km^3 (inset, Fig. 4). In this case the proportion of the volume inside the mapped isopachs is approximately 67% of the total giving a minimum deposited volume of 0.05 km^3 (Pyle 1995). We must also speculate the existence of a different distal decay line for the Upper Pumice eruption.

Tephrostratigraphy

Ash layers were sought with the cores collected during the 1975 *R.V. Trident* cruise 172 (1975) which are now kept at the University of Rhode Island. These cruise cores are positioned only to the south of Nisyros but several cores are in the direction of Aegean bottom currents (southeast trending) and tropospheric winds (east to southeast trending in the Aegean, Paterne 1985) away from Nisyros which should be ideally placed to record the deposition of tephra from Nisyros eruptions. The record of other large Aegean eruptions has also shown that distal tephra are invariably dispersed to the east and southeast (Federman & Carey 1980) primarily reflecting the prevailing tropospheric wind direction. The subaerial distribution of any distal Nisyros tephra, therefore, should also be controlled by this tropospheric wind direction and might plausibly be different to the dispersal directions implied by the proximal tephra onland. Tephra layers within the Aegean record can be recognized easily by inspection, since the dominant lithology is a light brown fine grained mud. Organic-rich sapropel layers provided obvious stratigraphic markers and were used to indicate the stratigraphic zonation of the layers within the alphabetic–numeric scheme proposed by Keller et al. (1978). Sapropels are labelled S1, S2 etc., following McCoy (1974) and Cita *et al.* (1977). These separate zones which are labelled Z, Y, X, W, V with increasing depth and ash layers were assigned to each zone in numeric order. The ash samples taken were $\geq 1\,\mathrm{cm}^3$ in volume. The minimum thickness of an ash layer recognizable as an eruption layer based on visual inspection could be regarded as 1 mm. In order to check for mixed ash occurrences, microscope smears were also taken at regular intervals in the closest cores to Nisyros, 172-25, -26 and -27. These were found to contain no ash at all in nearly all cases.

Distinct boundaries were usually seen between the ash and the surrounding clay and reworking or bioturbation, recognized by undulose or diffuse boundaries, was only seen in a few cases. For identification and correlation, the Munsell colour, thickness, state of preservation of the layer and mineralogy were checked before sections were prepared for electron microprobe analysis. An onland dataset of possible correlatives was compiled from glass concentrates of the silicic tephra sources from Nisyros: Panaghia Kyra, Lower Pumice, Upper Pumice; and the other silicic eruptions in the Aegean: Yali 1, 2, 3 and 4, the Kos Plateau Tuff and Santorini Minoan. 90% glass concentrates were obtained by crushing large ($>10\,\mathrm{cm}$) pumice clasts and separating the low density fractions on a Wilfley table. For other tephra layers where the only data available were whole rock XRF analyses rather than electron microprobe analysis only phenocryst-absent glass analyses were used where possible in order to be comparable with the ash layer analyses (Thunell *et al.* 1977).

EDS analytical method

The glass fractions were prepared for probe analysis by drying and mounting in an epoxy resin (Epotec). They were then ground to 40–50 μ using a Logitec and Al_2O_3 powder and polished using diamond paste. Energy dispersive spectrometry was performed on a CAMECA SX50 electron microprobe in the Department of Earth Sciences, Cambridge University. Standardization was carried out on the KN-18 comendite obsidian from Kenya (Neilsen & Sigurdsson 1981) in order to reduce Na loss and give accuracy and precision for 10 major elements. The result of changing beam current, beam diameter and count time is shown in Fig. 5. For a standard count time of 50 s at $\geq 30\,\mu\mathrm{m}$ beam diameter, 0.05 wt% Na is lost but under the conditions of a 2 nA beam current, $\geq 30\,\mu\mathrm{m}$ beam diameter and 20 s count time the analyses should require no correction. Na_2O and K_2O can be measured with accuracy of better than $\pm 2\%$ and a relative precision of $\pm 1\%$ or approximately 0.1 wt% oxide in the case of Na. The elements Al_2O_3, MnO, MgO, FeO, CaO have an accuracy of better than ± 0.005 wt% oxide. There is a higher statistical counting error involved with low count times but it is still not significantly greater than at

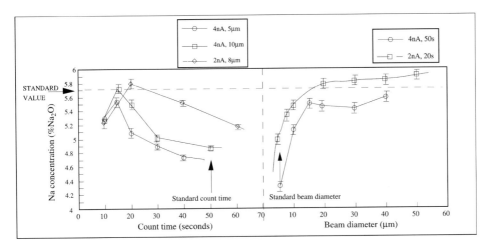

Fig. 5. Na$_2$O concentrations (wt%) determined by EDS analysis as a function of count time, beam current and beam diameter. Error bars represent 1 standard deviation based on counting statistics. For count times greater than 20-30 s, Na loss becomes appreciable. At beam diameters greater than 20 μm, a beam current of 2 nA and a count time of 20 s, Na$_2$O concentrations are comparable to those of the standard.

higher count times compared with the loss of Na. Repeated analysis indicated a relative precision of better than 1% under these conditions. For each analysis each shard was inspected for inhomogeneity and its size measured using a scanning electron microscope. Corrections for Na loss were only made where the glass shards were small. The number of analyses initially taken for each ash layer was a minimum of 20 and where inhomogeneity was found this was increased to 40.

Statistics

Principal components analysis was used to transform the dataset to a linear combination of variables that would account for most of the variance of the original dataset. It was performed using the commercially available statistics package Statview. The whole major element dataset (SiO$_2$, Al$_2$O$_3$, TiO$_2$, FeO, MnO, MgO, CaO, Na$_2$O and K$_2$O) was inputted in the raw data form with the totals renormalized to 100%. An orthotran/varimax transformation method was used to transform the data with extraction of 4 eigen vectors.

Compositional similarity coefficients were calculated between the EDS electron microprobe glass analyses of the ash layers and those of the possible source eruptions. These coefficients were calculated by dividing pairs of the glass analyses (ash layer to possible source eruption) element by element and then calculating the mean of these ratios (after Borchardt et al. 1971).

Results

Representative chemical analyses of the deposits from the principal regional silicic tephra sources: Nisyros, Santorini, Kos, and Yali are given in Table 1. All analyses are given with totals recalculated to 100% (anhydrous), with all Fe as FeO. Nisyros

Table 1. Aegean silicic tephra analyses

		SiO_2	TiO_2	Al_2O_3	FeO_T	MnO	MgO	CaO	Na_2O	K_2O	Total (No. of analyses for mean in brackets)	Author
NISYROS LOWER PUMICE (rhyolite)	Lowest SiO_2	74.61	0.26	12.95	1.53	0.08	bd	1.31	4.76	4.35	94.53	this work
	Highest SiO_2	76.53	0.18	13.04	1.47	bd	bd	0.87	4.13	3.829	3.27	
	MEAN	75.50	0.20	13.16	1.49	0.10	0.12	0.93	4.40	4.08	94.45 (72)	
	ST DEV	0.33	0.06	0.19	0.11	0.04	0.09	0.11	0.21	0.14		
NISYROS UPPER PUMICE (rhyolite)	Lowest SiO_2	76.21	0.30	12.40	1.39	0.03	0.16	0.94	4.06	4.47	93.08	this work
	Highest SiO_2	77.37	0.12	12.76	1.31	bd	bd	1.07	3.33	3.79	95.99	
	MEAN	75.75	0.20	12.68	1.35	0.03	0.02	1.06	3.72	4.09	95.40 (20)	
	ST DEV	0.86	0.11	0.29	0.10	0.06	0.09	0.32	0.31	0.11		
NISYROS PANAGHIA KYRA (zoned andesite to dacite)	Lowest SiO_2	57.44	0.38	17.91	5.81	0.06	6.57	5.63	1.40	0.41	98.72	this work
	Highest SiO_2	64.94	0.78	16.66	3.47	bd	1.02	.42	6.12	2.60	98.21	
SANTORINI MINOAN (rhyodacite)		70.76	0.31	13.70	2.06	bd	0.33	1.43	4.64	3.34	—	Keller 1980
SANTORINI CAPE RIVA (zoned andesite to rhyodacite)		73.80	0.28	14.09	1.94	bd	0.28	1.47	5.07	3.36	—	Druitt 1983
		63.50	0.70	16.40	6.00	bd	1.40	5.50	4.70	1.80	—	Druitt 1983
		71.60	0.40	14.50	2.60	bd	0.40	1.70	5.60	2.90	—	Druitt 1983
SANTORINI UPPER SCORIAE 2 (zoned andesite to dacite)		59.2	1.22	15.6	8.32	bd	2.68	6.43	4.42	1.67	100.0	Druit et al. 1989
		66.8	0.73	15.2	0.73	5.01	0.97	3.12	4.76	3.09	99.7	Druit et al. 1989
SANTORINI UPPER SCORIAE 1 (zoned andesite to dacite)		58.1	1.27	16.0	8.81	bd	2.90	6.62	4.76	1.25	99.7	Druit et al. 1989
		63.2	1.01	15.5	7.20	bd	1.58	4.48	4.65	1.96	99.8	Druit et al. 1989
SANTORINI VOUVOLOUS (zoned andesite to dacite)		59.30	1.29	14.83	9.91	0.16	1.43	4.80	4.33	1.95		Mellors 1988
		69.36	0.63	14.84	4.97	0.14	0.49	2.84	4.70	3.04		Mellors 1988
SANTORINI MIDDLE PUMICE (zoned dacite to rhyodacite)		64.9	0.30	15.80	5.70	bd	1.20	4.00	4.70	2.30	—	Druitt 1983
		71.00	0.30	14.30	3.60	bd	0.50	1.90	4.90	3.30	—	Druitt 1983
SANTORINI LOWER PUMICE (dacite)		67.80	0.25	14.22	2.02	bd	0.28	1.22	4.12	3.51	—	Keller 1980
		68.4	0.5	14.7	3.52	bd	0.83	2.17	6.21	3.48	100.00	Druit et al. 1989
KOS PLATEAU TUFF (rhyolite)	Lowest SiO_2	76.27	0.34	12.69	0.60	bd	bd	0.80	3.55	5.40	97.76	this work
	Highest SiO_2	77.60	0.15	12.41	0.40	bd	bd	0.51	3.74	5.109	8.09	
	MEAN	77.03	0.17	12.54	0.46	0.03	0.02	0.54	3.91	5.02	97.50 (14)	
	ST DEV	0.42	0.12	0.25	0.17	0.05	0.02	0.09	0.41	0.19		
YALI-1 (rhyolite)	Lowest SiO_2	76.51	0.14	12.60	1.22	0.01	0.1	0.80	4.09	4.59	94.28	this work
	Highest SiO_2	77.44	0.09	12.19	0.97	0.13	bd	0.73	3.55	4.70	94.64	
	MEAN	76.37	0.20	12.60	1.16	0.07	0.12	0.95	3.91	4.40	97.13 (35)	
	ST DEV	0.98	0.1	0.47	0.33	0.06	0.14	0.33	0.29	0.26		
YALI-2 (rhyodacite)	Lowest SiO_2	70.60	0.37	15.10	1.69	0.22	0.54	2.42	4.36	3.67	97.20	this work
	Highest SiO_2	72.46	0.44	14.66	1.63	bd	0.62	1.78	4.19	4.02	92.19	
	MEAN	71.12	0.42	14.63	1.96	0.06	0.62	2.11	4.59	3.80	95.58 (17)	
	ST DEV	1.38	0.14	0.26	0.61	0.08	0.40	0.46	0.42	0.20		
YALI-3 (rhyolite)	Lowest SiO_2	71.08	0.09	11.05	0.95	0.11	0.15	0.53	3.44	4.34	91.70	this work
	Highest SiO_2	76.24	0.24	11.97	0.87	bd	0.17	0.67	3.59	4.63	98.83	
	MEAN	1.55	0.10	0.58	0.20	0.07	0.22	0.64	0.25	0.24		
YALI-4 (rhyolite)		73.11	0.17	12.56	1.39	bd	1.00	3.39	4.21	3.33	—	Keller, 1980

bd, below detection

Table 2. Ash analyses of layers in Trident cores 172-25, 26 and 27

		SiO$_2$	TiO$_2$	Al$_2$O$_3$	FeO$_T$	MnO	MgO	CaO	Na$_2$O	K$_2$O	Total (no. of analyses in brackets)	Assigned tephra source
25PC (32.4–40.0 cm) ZONED	Lowest SiO$_2$	61.78	1.41	14.94	8.41	0.3	2.75	4.59	2.29	2.55	95.19	S-CR
	Highest SiO$_2$	73.37	0.3	13.58	2.13	0.05	0.17	1.63	4.82	3.84	93.75	
	MEAN	73.28	0.30	13.89	2.08	0.08	0.19	1.52	5.14	3.48	95.39(46)	
	ST DEV	1.13	0.06	0.27	0.24	0.09	0.26	0.46	0.12	0.08		
25PC (47.0–55.7 cm) ZONED	Lowest SiO$_2$	67.5	0.39	13.49	5.24	bd	0.89	1.38	5.98	3.41	97.50	S-MP?
	Highest SiO$_2$	74.61	0.20	13.95	2.06	0.06	0.1	1.21	5.34	3.50	91.93	
	MEAN	72.92	0.32	14.13	2.36	0.09	0.45	1.35	5.63	3.45	93.96(21)	
	ST DEV	1.50	0.08	0.25	0.71	0.08	0.87	0.11	0.49	0.16		
25PC (145 cm)	Lowest SiO$_2$	71.54	0.47	14.51	1.79	0.67	2.52	2.02	5.31	3.34	99.50	S-LP?
	Highest SiO$_2$	73.63	0.151	3.85	2.05	0.17	0.34	1.51	4.88	3.42	94.55	
	MEAN	72.64	0.34	13.86	2.07	0.23	0.74	1.72	5.00	3.45	94.92 (6)	
	ST DEV	0.70	0.12	0.55	0.15	0.22	0.88	0.29	0.31	0.11		
25PC (314–315 cm)	Lowest SiO$_2$	70.88	0.25	14.73	5.41	0.12	0.161	.25	4.14	3.92	95.52	K-PT
	Highest SiO$_2$	78.78	0.13	12.46	0.49	bd	bd	0.56	3.09	4.5	94.61	
	MEAN	75.99	0.02	12.62	0.38	0.03	0.45	0.33	3.39	5.87	92.67(5)	
	ST DEV	0.21	0.03	0.25	0.07	0.05	0.06	0.06	0.22	0.18		
26PC (0–4 cm)	Lowest SiO$_2$	71.23	0.2	11.58	1.08	bd	0.16	0.56	4.11	3.84	96.31S-M	S-CR
	Highest SiO$_2$	74.04	0.31	13.53	2.07	0.12	0.22	1.23	5.11	3.52	98.31	
	MEAN	73.30	0.31	13.73	1.95	0.1	0.24	1.38	4.94	3.47	96.34 (12)	
	ST DEV	0.76	0.08	0.71	0.29	0.16	0.12	0.28	0.31	0.15		
26PC (79–80 cm) ZONED	Lowest SiO$_2$	66.49	0.75	15.17	4.87	0.16	0.69	3.41	5.34	3.11	97.50	N-LP
	Highest SiO$_2$	69.50	0.72	14.72	4.22	0.05	0.43	2.18	5.78	3.24	94.24	
	MEAN	68.51	0.80	14.58	4.58	0.15	0.55	2.41	5.91	3.35	97.59 (13)	
	ST DEV	1.02	0.14	0.58	0.94	0.08	0.21	0.43	0.47	0.22		
26PC (133.1–133.4 cm)	Lowest SiO$_2$	72.08	0.02	12.42	0.84	0.19	0.26	0.97	4.14	3.62	98.07	
	Highest SiO$_2$	78.02	0.03	12.39	0.84	0.18	0.14	0.96	4.02	3.49	96.59	
	MEAN	77.48	0.11	12.65	0.77	0.11	0.14	0.97	4.29	3.71	96.32(31)	
	ST DEV	0.62	0.07	0.26	0.12	0.09	0.1	0.20	0.54	0.36		
26PC (405–409 cm)	Lowest SiO$_2$	75.50	0.19	12.21	0.65	bd	0.07	0.55	4.12	4.66	99.21	K-PT
	Highest SiO$_2$	77.29	0.06	12.65	0.50	bd	0.08	0.56	3.93	5.14	91.43	
	MEAN	77.04	0.13	12.56	0.55	0.02	0.06	0.56	3.95	4.87	95.13 (10)	
	ST DEV	0.60	0.06	0.14	0.21	0.05	0.06	0.07	0.22	0.28		
27PC (5–6 cm)	Lowest SiO$_2$	72.29	0.45	13.69	2.17	0.15	0.45	1.46	6.58	3.57	90.69	S-M
	Highest SiO$_2$	73.61	0.44	13.57	2.07	0.13	0.13	1.60	5.96	3.39	93.75	
	MEAN	72.95	0.45	13.63	2.12	0.14	0.29	1.53	6.27	3.48	92.67 (10)	
	ST DEV	0.94	0.01	0.08	0.07	0.01	0.23	0.10	0.44	0.12		
27PC (50.2–53.8 cm) ZONED	Lowest SiO$_2$	66.80	0.57	17.04	3.62	0.05	0.38	3.29	6.73	2.57	94.01	S-CR
	Highest SiO$_2$	69.83	0.82	13.48	3.98	0.07	0.17	2.33	5.83	3.35	92.99	
	MEAN	68.55	0.73	15.35	3.96	bd	0.42	2.49	6.32	3.12	93.54 (6)	
	ST DEV	0.99	0.09	0.93	0.26	bd	0.19	0.42	0.40	0.33		
27PC (289.5–290 cm) ZONED	Lowest SiO$_2$	53.98	1.02	17.10	8.93	0.14	6.27	7.82	4.79	0.91	94.83S-?	
	Highest SiO$_2$	63.11	1.55	13.48	8.85	0.28	1.36	4.05	5.20	2.69	95.22	
	MEAN	57.15	1.30	15.71	9.44	0.18	3.68	7.22	4.57	1.44	96.87 (11)	
	ST DEV	5.68	0.46	4.66	2.88	0.12	1.57	2.47	1.49	0.63		
27PC (305.5–306.0 cm) ZONED	Lowest SiO$_2$	56.98	1.32	15.58	9.08	0.03	3.79	4.80	4.85	2.57	99.0	2S-?
	Highest SiO$_2$	69.96	0.88	14.85	3.05	0.05	1.87	6.04	4.24	0.13	93.10	
	MEAN	60.14	1.09	16.17	5.80	0.08	2.38	5.48	5.46	3.06	92.34(10)	
	ST DEV	4.49	0.57	2.93	2.51	0.12	1.97	2.58	1.11	2.43		

Abbreviations: S-M (Santorini-Minoan), S-CR (Santorini-Cape Riva), S-MP (Santorini-Middle Pumice), S-LP (Santorini-Lower Pumice), N-LP (Nisyros-Lower Pumice), K-PT(Kos Plateau Tuff), bd, below detection. For the zoned layers, the mean is not a representative analysis for the layer but is an indicator of the compositional distribution within the zoned range.

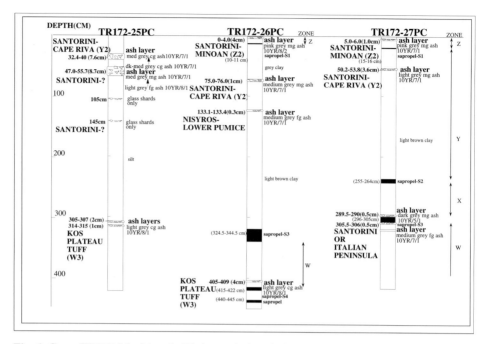

Fig. 6. Cores TR172-25, -26 and -27. A rough description of the colour and grain size of the ash is given followed by the Munsell: Hue/Value/Chroma. The grain size is defined as: fg = <30 μm, mg 30–50 μm and cg >50 μm.

Lower and Upper Pumice ash is distinguishable by inspection from the other sources of the Aegean due to its low Al_2O_3 content less than 13.0% and similar Na_2O and K_2O contents. The Lower Pumice always has Na_2O greater than K_2O but Na_2O ranges between 4.3–4.8 wt% and K_2O ranges between 4.0–4.3 wt%. The Upper Pumice always has Na_2O less than K_2O with Na_2O ranging between 3.3–4.2 wt% and K_2O between 3.8–4.5 wt%. The analyses reported here have higher Na contents than previous analyses of Nisyros pumice by on average 0.3 wt% (Keller 1980; Rehren 1988). The Nisyros Lower Pumice and Upper Pumice signature is similar to the Yali unit 1 signature. However the lack of Yali-1, -3 and -4 deposits on Nisyros and the reworked nature of the deposits on Yali suggest that of the Yali eruptions only Yali-2 would have been of large enough magnitude or the southerly dispersal direction necessary to appear in the Trident cores. The other Nisyros silicic sequence, the Panaghia Kyra series is distinguished by Na_2O values greater than K_2O, relatively high Al_2O_3 and CaO contents compared to the other tephra sources in the Aegean and it is compositionally zoned from andesite to dacite.

All the ash layers in the cores were analysed and compared to the source eruptions using the criteria mentioned above. These revealed no Nisyros tephra apart from core 172-26. Therefore, only the three cores closest to Nisyros – 172-25, -26 and -27 (Fig. 1) – are described here. The ash analyses from these cores are given in Table 2. The positions of the ash layers found in TR172-25, -26 and -27 are shown in Fig. 6. Their correlations are explained as follows.

TR172-25 (60 km SE of Nisyros). This core contains no sapropels making stratigraphic positioning of the tephra layers difficult. The first layer occurs at 40 cm depth and is 7.6 cm thick. It consists of colour zoned medium to dark grey, coarse-grained ash and it is compositionally zoned from dacite to rhyolite. It is thought to correspond to the Santorini–Cape Riva eruption. The Minoan layer, which is characterized by a pink-grey colour and an unzoned rhyodacite composition and, where present in other Trident cores, lies within the first 20 cm of sediment, is absent in this core. A layer of 8.7 cm thickness at 57 cm depth of medium-grained ash with light grey to grey colour zoning and a compositional range from dacite to rhyodacite is believed to represent another Santorini eruption. Of the eruptions predating the Cape Riva, possible eruptions with zoned compositions documented onland are the Vourvolous or the Middle Pumice (Druitt et al. 1989). Of these two eruptions, only the Middle Pumice has been postulated to occur in the Aegean tephrostratigraphic record (Federman & Carey 1980). A thin layer of glass shards at 145 cm depth has unzoned rhyodacite composition and a chemical signature typical of Santorini magmas. The source of this eruption could be the Santorini Lower Pumice. Two eruptions at 305–307 cm and 314–315 cm depth of light grey coarse-grained ash with rhyolite composition are thought to correlate with the Kos Plateau Tuff eruption. None of the tephra analyses within 172-25 have the chemical characteristics of Nisyros tephra, which excludes Nisyros as a possible source for any of these layers.

TR172-26 (75 km SW of Nisyros). A layer of 4 cm thickness at 4 cm depth of pink-grey medium-grained ash and an unzoned rhyodacite composition corresponds to the Santorini Minoan eruption. Below sapropel S1 a layer of medium-grained grey ash of 1 cm thickness at 80 cm depth with a zoned dacite composition is believed to correspond to the Santorini Cape Riva eruption (Y2). A layer of fine-grained grey ash with unzoned rhyolite composition of 3 mm thickness at 133.4 cm depth is thought to correspond to a Nisyros eruption. It has the signature of the Nisyros Lower and Upper Pumice tephra with Al_2O_3 content <13.0% and similar Na_2O (4.1–4.5%) and K_2O (3.9–4.2%) ratios. The similarity coefficients between this layer and the silicic tephra sources of the Aegean confirm that this layer is most similar to the Lower Pumice tephra (Table 3). The similarity coefficients within the Nisyros layer itself are greater than 99% and the standard deviations are <0.7 which indicates that the layer represents an unzoned eruption from a single source (Table 4). In a plot of eigen vector 1 and eigen vector 2 of the tephra sources and the postulated Nisyros layer, the signature of ash layer lies between the Nisyros Lower Pumice and the Yali-2 pumice but it lies closest to the Nisyros Lower Pumice (Fig. 7). A 4 cm thick layer of light grey coarse-grained rhyolitic ash at 409 cm depth is correlated with the Kos Plateau Tuff eruption (W3).

TR172-27 (85 km SW of Nisyros). The first layer occurs at 6 cm depth of 1 cm thickness and consists of the pink-grey medium-grained ash with unzoned rhyodacite composition characteristic of the Santorini Minoan eruption (Z2). Below sapropel S1, within the Y zone, a 3.6 cm thick layer at 54 cm depth of light grey medium grained ash and a zoned dacite composition is correlated with the Santorini Cape Riva eruption (Y2). A 0.5 cm thick layer of dark grey medium grained ash at 289.5–290 cm depth below sapropel S2 has a zoned andesite to dacite composition. Another 0.5 cm thick layer of dark grey fine-grained ash with a similar

Table 3. *Similarity coefficients of ash layer 26PC(133.1–133.4 cm) against mean analyses of silicic glass components of the most probable silicic tephra sources for the eastern Aegean*

	SiO$_2$	TiO$_2$	Al$_2$O$_3$	FeO$_T$	CaO	Na$_2$O	K$_2$O	Mean
Nisyros Lower Pumice	97.44	45.83	96.12	51.68	96.00	97.5	90.93	**87.31**
Nisyros Upper Pumice	98.77	55.00	99.76	57.04	91.51	87.00	90.71	**82.83**
Nisyros Panaghia Kyra (dacite)	81.20	11.34	74.11	17.15	25.94	69.53	68.73	**49.71**
Nisyros Panaghia Kyra (rhyodacite)	89.78	9.65	92.27	16.18	36.06	95.55	92.72	**61.74**
Santorini Minoaan	93.24	39.29	91.07	40.74	68.31	85.46	88.14	**72.32**
Santorini Cape Riva (andesite)	81.96	15.71	77.13	12.83	17.64	91.28	48.52	**49.30**
Santorini Cape Riva (dacite)	92.41	27.50	86.64	25.67	57.06	76.61	78.17	**63.44**
Santorini Upper Scoriae (silicic andesite)	83.18	12.64	81.4	11.63	23.49	90.89	50.67	**50.56**
Santorini Upper Scoriae (andesite)	77.54	8.40	78.91	9.28	16.67	96.84	39.62	**46.75**
Santorini Middle Plumice (andesite)	82.73	13.75	80.06	14.61	27.79	93.87	56.06	**52.70**
Santorini Middle Plumice (daacite)	91.64	36.67	88.46	21.39	51.05	87.55	88.95	**66.53**
Santorini Lower Pumice	87.51	44.00	88.96	38.12	79.51	96.00	94.61	**75.53**
Kos Plateau Tuff	99.42	64.71	99.13	59.74	55.67	91.14	73.90	**77.67**
Yali-1	98,57	55.00	99.60	66.38	98.00	91.00	84.32	**84.81**
Yali-2	91.79	26.19	86.47	39.29	45.97	93.46	97.63	**68.69**
Yali-3	97.79	57.89	98.52	52.03	92.38	92.54	82.35	**82.35**
Yali-4	94.50	64.71	99.29	55.40	28.61	98.14	76.95	**76.95**

Table 4. *Similarity coefficients within the 'Nisyros' ash layer itself*

	26PC (133.1–133.4 cm)	26PC (133.1–133.4 cm)	26PC (133.1–133.4 cm)	26PC (133.1–133.4 cm)	26PC (133.1–133.4 cm)	26PC (133.1–133.4 cm)
26PC (133.1–133.4 cm)	100.00	–	–	–	–	–
26PC (133.1–133.4 cm)	99.78	100.00	–	–	–	–
26PC (133.1–133.4 cm)	99.86	99.92	100.00	–	–	–
26PC (133.1–133.4 cm)	99.92	99.78	99.85	100.00	–	–
26PC (133.1–133.4 cm)	99.86	99.93	1000.00	99.83	100.00	

zoned andesite to dacite composition occurs beneath sapropel S3 at 305.5–306 cm depth. The zoned composition suggests that Santorini may be the origin for these layers but the basaltic andesite composition of the low SiO$_2$ end member excludes most of the documented Santorini eruptions unless it represents an eruption which left no tephra record proximally. Alternative sources for these layers are considered to be the Italian Peninsula calc-alkaline centres of Etna or the Aeolian Islands.

Discussion

The Nisyros Lower and Upper Pumice eruptions do not appear to have left a deep sea tephra record of widespread extent. This agrees with a sub-Plinian classification deduced from isopach data and pumice and lithic isopleth data onland. It also indicates that the volume estimates for these eruptions would be small (< 1 km^3 bulk DRE magma). This agrees with a hypothesis that the extrusion of the Nikia Rhyolite would have caused a greater volume loss in the caldera formation and suggests that caldera collapse on the island was incremental. The Nisyros ash is not found in the closest core to the island, but in the postulated downwind direction based on the ellipticity of the isopleth/isopach contours. This ash layer is also found further from the island than that predicted from just the proximal part of the

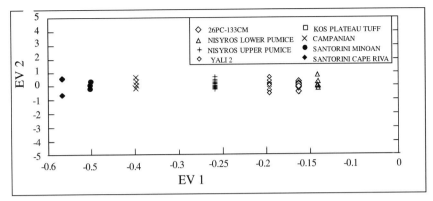

Fig. 7. Eigen vector 1 plotted against eigen vector 2 for the known Aegean tephra sources and the ash in TR 172-26PC (133 cm).

isopach data if redistribution by bottom currents is ruled out and is consistent with the observation that many deposits thin at different rates. The elliptical, NE–SW directed, isopach orientations could explain why there are no findings of Nisyros Lower and Upper Pumice tephra on Pakia and Perigusa islands, 4 km to the west (Limburg & Varekamp 1991) or Tilos, 18 km to the southeast (Rehren 1988). No discrete ash layer corresponding to the Upper Pumice eruption was found. This would agree with the northerly dispersal direction of this eruption indicated by the isopach data and indicates that this eruption was of smaller dispersal than the Lower Pumice.

Only a rough guess, based on extrapolation of a uniform sedimentation rate through the 172-26 core, may be made with respect to the Lower Pumice age. The sediment thickness between the *c.* 3500 a Santorini Minoan eruption and the 21 ka Santorini Cape Riva eruption suggests an average sedimentation rate of *c.* 4 cm ka^{-1} and an age of 35 ka for the 133 cm depth of the Lower Pumice. Its depth relative to the other cores in the vicinity is similar to the Yali-2 layer whose depths range between 100 and 130 cm depths is dated at 31 ka by 18O stratigraphy (Federman & Carey 1980), or the Campanian Y-5 layer whose depths range between 130 and 150 cm and is dated at 38 ka, (Thunell *et al.* 1979). If sedimentation rates are comparable from core to core, which the positions of the younger Cape Riva and older Kos Plateau Tuff layers suggest they are, this also suggests that the Lower Pumice is of the same order of age as these eruptions. The onland stratigraphy of Nisyros also places the Lower Pumice eruption between the mid-stratocone dacite lavas dated at 38 ka (Keller *et al.* 1989) and the Yali-2 pumice found inside the Nisyros caldera at the base of the northern wall and overlying the Upper Pumice at Pali promontory (Rehren 1988).

Ash has previously been assigned to the Nisyros 'caldera phase' (Upper Pumice; Keller 1980) by Vinci (1983, 1985). It was found in the *R/V Bannock* cruise core (1981) MC27 (latitude 35°36′ 0 and longitude 26°49′ 0) in a 7 cm thick layer at 307 cm depth. This core lies 120 km to the south of Nisyros. Vinci's correlation is important because it has been used to estimate an age of 24 ka for the Upper Pumice and an assumed upper limit of caldera formation. It also formed the basis for the large DRE volume estimate inferred for the Upper Pumice eruption assuming that the layer represented a primary fall deposit (2–3 km^3, Limburg & Varekamp 1991). The only way that the discovery can fit in with the hypotheses here is if this layer is assigned to the Kos Plateau Tuff eruption. The depth (300 cm) and thickness (7 cm) correspond to the other analyses of the Kos Plateau Tuff found throughout the southeast Aegean. The chemical composition of the MC27 layer is as close to the Kos Plateau Tuff as it is to the Upper Pumice. The Na$_2$O and K$_2$O contents are closely similar to the Kos Plateau tuff (Na$_2$O = 3.45, K$_2$O = 4.66) and the other elements are similar for the two eruptions. Similarity coefficients are equal between the ash layer and the two eruptions and principal component analysis places the layer in between the two eruptions. Assigning this layer to the KPT also produces a more reasonable sedimentation rate for core MC27 since a sedimentation rate of 13 cm ka^{-1} would be necessary to obtain the age of 24 ka for a 300 cm depth layer which is 6 times higher than the average sedimentation rate in the southeast Aegean. If it is the Upper Pumice it raises the problem of the lack of similar sediments in other cores in the region and the absence of exposures of Upper Pumice on islands closer to Nisyros e.g. Karpathos near the core, Tilos, Kandeliusa and the volcanic sub-islets around the island since from evidence of other distal fall deposits a 7 cm distal tephra fall layer at 120 km from source, would imply a regionally correlatable deposit over the scale of at least 10^5 km^2. The thickness of ash at this distance from the source would also indicate dispersal from a Plinian column which contradicts all the onland evidence for the Upper Pumice eruption. The possibility that it represents an Upper Pumice ignimbrite or co-ignimbrite ash cloud is also considered unlikely since the Upper Pumice unit D was directed eastwards, and prevailing winds and marine currents would have transported airborne and seaborne ash southeastwards (Paterne 1985).

Conclusions

Nisyros recent caldera-forming eruptions were small sub-Plinian events capable of contributing to the deep sea tephra record but not on a laterally correlatable scale. The Lower Pumice was the larger eruption indicated by both the tephra distribution onland and its finding in the deep sea record. Both eruptions have a NE–SW distribution direction but the Lower Pumice occurred with a vent on the eastern side of the ancestral cone and the Upper Pumice with a vent in a more northerly central position. The bulk DRE of the tephra deposits from each eruption is likely to have been less than 1 km^3 magma in each case.

I thank Dr S. Carey at the University of Rhode Island for permission to look at the Trident cores and obtain the samples. This work was carried out during the tenure of a NERC PhD studentship.

References

BORCHARDT, G. A., HARWARD, M. E. & SCHMIDT, R. A. 1971. Correlation of volcanic ash deposits by activation analysis of glass separates. *Quaternary Research*, **1**, 247–260.

CITA, M. B., VERGNAUD-GRAZZINI, C., ROBERT, C., CHAMLEY, H., CLARANFI, N. & D'ONOFRIO, S. 1977. Paleoclimatic record of a long deep-sea core from the eastern Mediterranean. *Quaternary Research*, **8**, 205–235.

DI PAOLA, G. M. 1974. Volcanology and petrology of Nisyros island (Dodecanese, Greece).

Bulletin of Volcanology, **38**, 944–987.

DRUITT, T. H. 1983. *Explosive volcanism on Santorini, Greece*. PhD thesis, Cambridge University.

——, MELLORS, R. A., PYLE, D. M. & SPARKS, R. S. J. 1989. Explosive volcanism on Santorini, Greece. *Geological Magazine*, **126**, 95–126.

FEDERMAN, A. N. & CAREY, S. N. 1980. Electron microprobe correlation of tephra layers from eastern Mediterranean abyssal sediments and the island of Santorini. *Quaternary Research*, **13**, 160–171.

FERRARA, G., FYTIKAS, N., GUILIANI, O. & MARINELLI, G. 1980. Age of the formation of the Aegean active volcanic arc. *In:* DOUMAS, C. (ed.) *Thera and the Aegean World, II*, Thera and the Aegean World, London, Athens, 37–41.

FIERSTEIN, J. & NATHENSON, M. 1992. Another look at the calculation of tephra fall volumes. *Bulletin of Volcanology*, **54**, 156–167.

FRANCALANCI, L., VAREKAMP, J. C., VOUGIOKALAKIS, G., DEFANT, M. J., INNOCENTI, F. & MANETTI, P. 1995. Crystal retention, fractionation and crustal assimilation in a convecting magma chamber, Nisyros volcano, Greece. *Bulletin of Volcanology*, **56**, 601–620.

FYTIKAS, M., GIULIANI, O., INNOCENTI, F., MARINELLI, G., & MAZZUOLI, R. 1976. Geochronological data on recent magmatism of the Aegean sea. *Tectonophysics*, **31**, 29–34.

——, INNOCENTI, F., KOLIOS, N, MANETTI, P., MAZZUOLI, R. *et al.* 1986. Volcanology and petrology of volcanic products from the Island of Milos and neighbouring islets. *Journal of Volcanology and Geothermal Research*, **28**, 297–317.

KELLER, J. 1971. The major volcanic events in recent eastern Mediterranean volcanism and their bearing on the problem of Santorini ash-layers. International Science Congress, Volcano of Thera, 1st, 1969, *Acta, Greek Archaeological Service*, 152–167.

—— 1980. Prehistoric pumice tephra on Aegean islands, *In:* DOUMAS, C. (ed.) *Thera and the Aegean World II*, Thera and the Aegean World, Athens, 49–56.

—— 1981. Quaternary tephrochronology in the Mediterranean region. *In:* SELF, S. & SPARKS, R. S. J. (eds) *Tephra Studies*. Reidel, Dordrecht 227–244.

——, GILLOT, P. Y., REHREN, T. H. & STADLBAUER, E. 1989. Chronostratigraphic data for the volcanism in the eastern Hellenic arc; Nisyros and Kos. *Terra Abstracts*, **0S06-26**, 354.

——, REHREN, T. H. & STADLBAUER, E. 1990. Explosive volcanism in the Hellenic arc. *In:* DOUMAS, C. (ed.) *Thera and the Aegean World, III*, Vol. 2, Thera and the Aegean World, Athens, 13–26

——, RYAN, F., NINKOVITCH, D. & ALTHERR. 1978. Explosive volcanic activity in the Mediterranean over the past 200,000 yr recorded in deep sea sediments. *Geological Society of America Bulletin*, **89**, 591–604.

LIMBURG, E. M. & VAREKAMP, J. C. 1991. Young pumice deposits on Nisyros, Greece. *Bulletin of Volcanology*, **53**, 68–77.

LINEWEAVER, J. L. 1962. Oxygen outgassing caused by electron bombardment of glass. *Journal of Applied Physics*, **34**, 1786–1791.

MARINI, L., PRINCIPE, C., CHIODINI, G., CIONI, R., FYTIKAS, M. & MARINELLI, G. 1993. Hydrothermal eruptions of Nisyros, Dodecanese, Greece. Past event and present hazard. *Journal of Volcanology and Geothermal Research*, **56**, 71–94.

MCCOY, F. W. 1974. *Late Quaternary sedimentation in the eastern Mediterranean sea*. PhD thesis, Harvard University, Cambridge, Massachusetts.

MELLORS, R. 1988. *Explosive volcanism on Santorini, Greece and Mt St. Helens*. PhD thesis, Cambridge University.

NEILSON, C. H. & SIGURDSSON, H. 1981. Quantative methods for electron microprobe analysis of sodium in natural and synthetic glasses. *American Mineralogist*, **66**, 547–552.

PAPAZACHOS, B. C. & COMNINAKIS, P. E. 1971. Geophysical and tectonic features of the Aegean arc. *Journal of Geophysical Research*, **35**, 8517–8533.

PATERNE, M. 1985. [*Reconstruction of the volcanic activity from the volcanoes of southern Italy using tephrochronology*]. PhD thesis, Paris-Sud University, France [in French].

PE, G. G. 1973. Petrology and geochemistry of volcanic rocks of Aegina, south Aegean Arc, Greece. *Bulletin of Volcanology*, **37**, 491–514.

Pe-piper, G. G., Piper, D. J. W. & Reynolds. P. H. 1983. Paleomagnetic stratigraphy and radiomatric dating of the Pliocene volcanic rocks of Aegina, Greece. *Bulletin of Volcanology*, **46**, 1–7.

Pyle, D. M. 1989. The thickness, volume and grain size of tephra fall deposits. *Bulletin of Volcanology*, **51**, 1–15.

—— 1995. Assessing uncertainty in the volume of tephra fall deposits. *Journal of Volcanology and Geothermal Research*, **69**, 379–382.

Rehren, T. H. 1988. [*Geochemistry and petrology of Nisyros*]. PhD thesis, Freiburg University [in German].

Seidenkrantz, M. & Friedrich, W. L. 1993. Santorini, part of the Hellenic Arc: Age of the earliest volcanism documented by foraminfera. *Bulletin of the Geological Society of Greece*, **28**(3), 99–115.

Stadlbauer, E. 1988. [*Volcanological and geochemical analysis of a young ignimbrite: the Kos Plateau Tuff (SE Aegean)*]. PhD thesis, Freiburg University [in German].

Thorarinsson, S. 1954. [The eruptions of Hekla 1947–48. II-3. The tephra fall from Hekla]. *Vis. Islendinga*, Reykjavik, 68 [in Icelandic]

Thunell, R., Federman, A., Sparks, R. S. J. & Williams, D. 1979. The age, origin and volcanological significance of the Y-5 ash layer in the Mediterranean. *Quaternary Research*, **12**, 241–253.

——, Williams, D., Federman, A. & Sparks, R. S. J. 1977. Late Quaternary tephrochronology of eastern Mediterranean sediments. *Geological Society of America Abstracts with programs*, **9**, 1200.

Vinci, A. 1983. A new ash layer 'Nisyros Layer' in the Aegean sea sediments. *Bolletino di Oceanologia Teorica ed Applicata*, **1**, 341–343.

—— 1985. Distribution and chemical composition of tephra layers from eastern Mediterranean abyssal sediments. *Marine Geology*, **64**, 143–155.

Vougiokalakis, G. 1992. [Volcanic stratigraphy and evolution of Nisyros]. *Bulletin of the Geological Society of Greece*, **28**(2), 239–258 [in Greek].

Watkins, N. D., Sparks, R. S. J., Sigurdsson, H., Huang, T. C., Federman, A., Carey, S. & Ninkovich, D. 1978. Volume and extent of the Minoan tephra from Santorini volcano: new evidence from deep sea sediment cores. *Nature*, **271**, 122–126.

Eruptive and seismic activity at Etna Volcano (Italy) between 1977 and 1991

S. VINCIGUERRA,[1,3] S. GAROZZO,[1,4] A. MONTALTO[2] & G. PATANÈ[1,4]

[1] *Istituto di Geologia e Geofisica, Università di Catania, Italy*
[2] *Dipartimento di Scienze della Terra, Università di Pisa, Italy*
[3] *Benfield Greig Hazard Research Centre, Department of Geological Sciences, University College London, Gower St, London WC1E 6BT, UK*
[4] *Present address: Dipartimento di Scienze Geologiche, Università di Catania, Catania, Italy*

> **Abstract:** A review of volcanic and seismic activity at Etna (Sicily, Italy) between 1977–1991 is presented with the purpose of looking for possible relations between the main phases of seismic energy release and the principal eruptive episodes. Flank eruptions are normally preceded by both deep and shallow seismic swarms ($0 < h < 20$ km) which accompany the opening of fracture systems some hours before the eruption onset. Conversely, summit eruptions usually take place without any immediate change in the seismic patterns, even if moderate seismic activity some months before the onset suggests that magma uprises gradually and might rest within shallow reservoirs during ascent. It is worth noting that in some cases (1981, 1983, 1984 and 1986) significant episodes of seismic energy release occurred during the late phases of the eruptions. The analysis of the temporal evolution of the coefficient b of the magnitude–frequency relation suggests that the shallow seismicity of the eastern flank reflects the existence of a high level of geodynamic activity, probably associated with both regional tectonic forces and seaward sliding of this unbuttressed sector of the volcano.

This paper considers the relationship between volcanic and seismic activity at Etna (Sicily, Italy) between 1977 and 1991. The main objective is to identify any correlations between two phenomenologies, by taking into account that eruptive events consist of both summit and flank eruptions and also by considering that significant episodes of seismic energy release occurred with no apparent temporal relation with volcanic activity. A detailed analysis is carried out of the spatial and temporal variations of the b coefficient of the magnitude–frequency relationship.

Seismic energy releases at Etna before, during and after an eruption can give useful signs about processes causing eruptive phenomena. Sharp *et al.* (1981) find a significant relationship between flank eruptions and earthquakes in the region immediately around the volcano. Moreover, summit eruptions are found to occur before flank eruptions which are not preceded by earthquakes. Scarpa *et al.* (1983) stated that seismological monitoring is a succesful method for predicting lateral eruptions of the volcano, since they are related, albeit in a complex way, to the seismic episodes. Nercessian *et al.* (1991) concentrated their attention on earthquakes occurring at the end of eruptive phases and concluded that large magnitude seismic activity around Etna is closely related to lateral eruptive episodes, probably

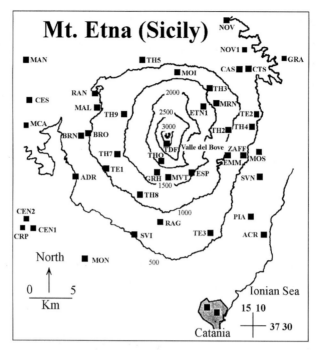

Fig. 1. Mt. Etna map: location and names of the seismic stations (filled squares).

due to a redistribution of stresses and pressures in the heterogeneous mass of the edifice, while Gasperini *et al.* (1990) suggested that there is no relationship between seismic activity and eruptions.

Data analysis

The dataset comprise about 6000 earthquakes recorded between 1 January 1977 and 31 December 1991 by means of the permanent network of the University of Catania, and the temporary networks run by the University of Cambridge (1982) and by ARGOS (1983) (Fig. 1). We first carried out the computation of the strain release associated with crustal earthquake activity in the framework of the eruptive events occurring over the time interval investigated. Strain release plots were calculated after the evaluation of the energy content of earthquakes by means of the Richter relationship (1958):

$$\text{Log}E = 9.9 + 1.9M_L - 0.024M_L^2$$

We computed separate plots for the whole Etnean area (Fig. 2a) and separately for the eastern and western flanks (Fig. 2b) in order to recognize possible correlations between the behaviour of the storage and release of seismic energy and eruptive phenomena (both effusive and explosive) over the considered period.

Figure 2a shows the global strain release. It is important to note that in the Etnean area the energy releases occurred in moderate and successive episodes, mainly involving earthquakes with magnitudes in the range 2.6–3.9. Careful observation permitted the differentiation of 3 trends. Between 1977 and early 1983 Etna was

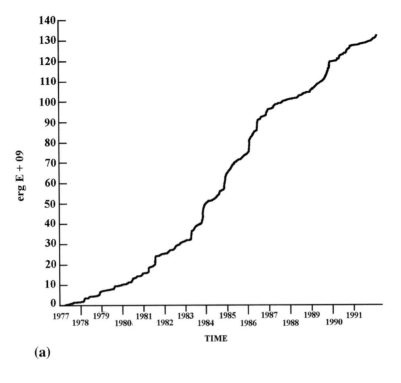

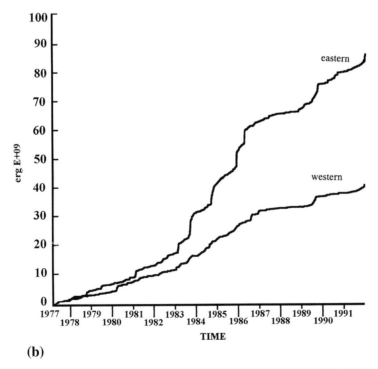

Fig. 2. Cumulative seismic energy release of (**a**) the Etnean area during 1977–1991 and (**b**) for the eastern and western flank.

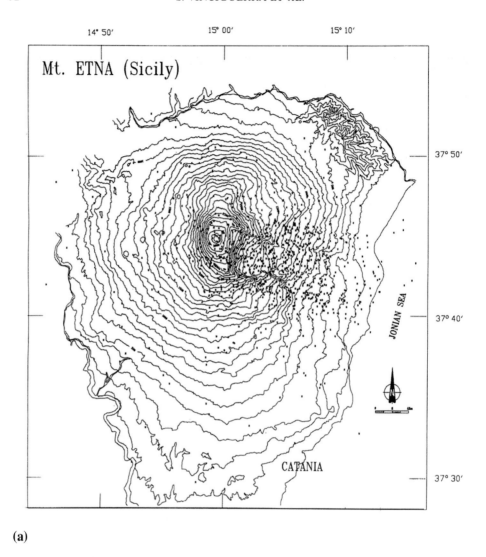

(a)

Fig. 3. (**a**) Map of the area investigated and epicentral distribution (only events with epicentral standard errors less than 3 km are represented). (**b**) Schematic map of 1971–1991 lava flows: subterminal (grey) and flank eruptions (black) (from Azzaro & Neri 1992).

characterized by quite low energy release in comparison with the successive period, which lasted until the end of 1986. Afterwards, a different pattern was recorded, probably associated with a modification in the behaviour of the storage of elastic deformation.

The seismic behaviour of the two flanks seems quite different. In fact over the period considered, the western flank is characterized by gradual moderate energy releases, in particular during the period between 1987 and 1991. In contrast, the eastern flank is characterized by considerable energy releases, mainly related to some powerful episodes which took place between 1983 and 1986. This is apparently supported by the spatial distribution of the epicentres (Fig. 3a) which are

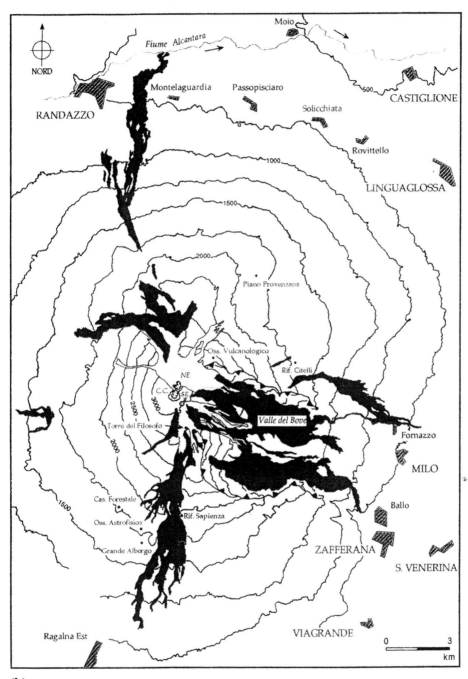

(b)

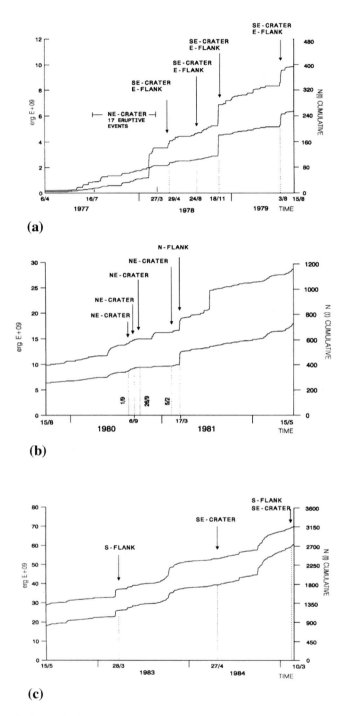

Fig. 4. Cumulative seismic energy release (complete line) and cumulative number of earthquakes (dashed line) between (**a**) 6 April 1977–15 August 1979, (**b**) 15 August 1979–15 May 1982, (**c**) 15 May 1982–10 March 1985, (**d**) 10 March 1985–14 December 1987, (**e**) 14 December 1987–25 August 1990, (**f**) 25 August 1990–31 December 1991.

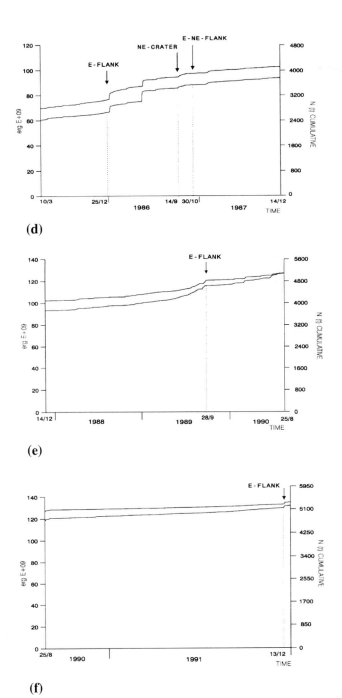

predominantly concentrated within the 'Valle del Bove'.

In order to appreciate better the possible relationships between the occurrence of seismic and volcanic activity, we separated the general curves presented above into individual plots including discrete phases of the eruptive chronicle for the period investigated.

Volcanic and seismic activity associated with the eruptive phases

In this section we report the seismic and volcanic activity which occurred during the period 1977–1991 (Fig. 3a,b). Information about volcanic activity is taken from Azzaro & Neri (1992), the SEAN Bulletin (1978, 1979, 1981, 1984, 1985, 1986, 1989, 1991) and Tanguy & Patanè (1984).

(i) The July 1977–March 1978 subterminal activity. Strombolian activity occurred in the NE crater, with 17 effusive episodes of very short duration on the NW flank of the crater. Lava flows affected the northern and the northwestern flanks of the volcano, as well as the Valle del Bove. The episodes occurred on: 16 July 1977 (6 days), 5 August 1977 (2 days), 14 August 1977 (1 day), 2 November 1977 (2 days), 7 November 1977 (2 days), 22 November 1977 (1day), 25 November 1977 (1 day), 6 December 1977 (1 day), 10 December 1977 (3 days), 18 December 1977 (1 day), 24 December 1977 (1 day), 29 December 1977 (1 day), 2 January 1978 (1 day), 5 January 1978 (1 day), 7 January 1978 (1 day), 25 March 1978 (2 days), 27 March 1978 (1 day). No particular seismic activity was related with 15 of the 17 eruptive episodes (Fig. 4a). Earthquakes occurred predominantly during November and January 1978 with weak events (M <2.5). Contrastingly, an appreciable energy release preceded the last two eruptive episodes occurring on 25 and 27 March respectively. This was due to a seismic swarm of several events that occurred on 23 and 24 February. This swarm included earthquakes with magnitudes generally between 2.5 and 3.0. Moreover, other events occurred at the beginning of March, including an earthquake on the 11 March with a magnitude of 3.8.

(ii) The April–June 1978 flank eruption. The onset of the eruption occurred on 29 April 1978 along an eruptive fracture located in the SE flank, which, two days later, crossed the western rim of the Valle del Bove. On 1 May lava was extruded from 3 new boccas, located along the same fracture. Only one bocca was active during the entire eruption, whereas the other two stopped, respectively, on 1 and 7 May. The lava flows were accompanied by strong phreato magmatic explosions and gas emission under high pressure. The eruption ended on 5 June 1978. No seismic activity occurred before the beginning of this eruption (Fig. 4a). Only a moderate energy release accompanied its development, due to a few earthquakes that occurred in May, particularly during the morning of the 17 May with magnitudes between 2.5 and 3.2.

(iii) The August 1978 flank eruption. The eruption started on 25 August 1978 at the SE crater and affected the Valle del Bove. It was preceded by ejection of spatter, bombs and ash from one of the 1971 eruption craters, at 3000 m altitude on the east flank of the SE crater, and lava overflows from the SE crater itself. On the 26 August a second vent opened on the western rim of the Valle del Bove. The explosive

activity that had started on 24 August began to decrease on 27 August, but 7 more vents opened along a fracture downhill, around the 1908 craters. By 29 August, the number of active vents had decreased to 4 with a notable diminution of lava effusion, and explosions had ended. At the same time another vent opened near the north rim of Valle del Bove, along the 1971 eruptive fracture. The eruption ended on 30 August 1978. As in the previous case, no appreciable seismic activity occurred before the onset of the eruption (Fig. 4a). Three earthquakes with magnitude around 3.0 occurred during August generating a moderate energy release that preceded the start of the eruption. An event of M = 3.2 occurred during the eruption on 26 August.

(iv) The November 1978 flank eruption. The eruption started on 11 November 1978 and ended on 30 November 1978. Initial activity consisted of ejection of ash and wallrock from one of the spatter cones formed along the fracture from the previous eruption. On 25 November two boccas opened in the wall of the Valle del Bove. Three other new boccas opened on 27 November at lower altitude, producing lava flows.

Another vent produced a lava flow affecting the Valle del Bove towards Milo. A powerful seismic swarm occurred on 26–28 November 1978, associated with the opening of three eruptive vents. On this occasion, several earthquakes with magnitudes between 2.0 and 3.0 occurred, affecting mainly the summit area and the Valle del Bove. Such activity resulted in a significant energy release (Fig. 4a).

(v) The August 1979 flank eruption. The eruption was preceded in early July by strong Strombolian activity at the Chasm. On 16 July the beginning of Strombolian activity was observed at the SE crater. On 27 July a collapse at Bocca Nuova was heard and billowing clouds of dust were emitted. The eruption began on 3 August, with eruptive systems opening in four days. The final fracture formed on 6 August near the Rifugio Citelli. The eruption ended on 9 August, after which Strombolian activity occurred during the night between 1–2 September, with a collapse of part of the wall of the Bocca Nuova and small explosions from the Chasm, the following day. On 12 September a strong explosion from the Bocca Nuova killed 9 tourists and injured another 23. A seismic swarm occurred on 30 July with a few earthquakes included between magnitudes 2.6 and 3.3. A more appreciable swarm occurred on 3–4 August accompaning the start of the eruption. Magnitudes were generally between 2.0 and 3.0. The localization of the swarm was similar to that of the swarm preceding the November 1978 eruption, and was also characterized by a significant energy release (Fig. 4a).

(vi) The September 1980 subterminal activity. This eruptive period is characterized by continued Strombolian activity at the NE crater and lava flows developed on the NW flank of the crater on 1–2, 6–7 and 26 September, respectively. From June to August 1980 appreciable seismic activity occurred (Fig. 4b). Nevertheless, there were no particular temporal clusterings and the events seemed to be spread out over the period considered, resulting in a gradually increasing energy release. Magnitudes were generally between 2.5 and 3.2 and the epicentres (Schick *et al.* 1982), were mainly located along the NW–SE tectonic trend. Some earthquakes occurred at the end of August provoking damage to some villages located on the southern flank,

while others occurred during September coinciding with the eruptive episodes. It is important to stress that the epicentres were scattered on the volcanic edifice. Episodes of creep were also observed, particularly on the low eastern flank.

(vii) The February 1981 subterminal activity. After a period characterized by ash emission, between the end of January and the beginning of February 1981, stronger activity followed with intense explosions on the evening of 5 February. Lava flowed through a breach in the W-to-NW side of the NE crater cone, and formed 3 lobes that moved W, NW and N, and covered the upper NW slope of the volcano. The north lobe, the largest, travelled about 2 km to about 2600 m elevation where it had a 1.2 km front. The eruption had a duration of only two days. Some earthquakes with magnitudes between 3.0 and 3.7 occurred at the end of November and in early December 1980, giving an appreciable energy release (Fig. 4b). An isolated event also occurred on 8 January at Piano Provenzana, probably linked with a displacement of the Pernicana fault, causing damage to the local buildings. No seismic activity occurred at the time of the eruption.

(viii) The March 1981 flank eruption. The eruption of 17–23 March extruded lava from several fissures which opened over a period of 33 hours on the NNW flank, forming an eruptive system with a length of about 8 km. The main flow reached the Alcantara River bed on 19 March at 1100 m altitude, while the flows extruded from the fissure between 1235 m and 1140 m altitude continued to advance slowly. The start of the eruption was accompanied by swarms of several hundred earthquakes on 16 and 17 March, linked to the opening of the fracture and magma intrusion (Fig. 4b). Magnitudes were between 2.0 and 3.0. A number of events accompanied the six days of eruption, with magnitudes generally between 2.5 and 3.0. Scarpa *et al.* (1983) noted that the distribution of these epicentres relative to those of earthquakes taking place at the time of the eruptive events of July–December 1977, November 1978, September 1980 and March 1981, seemed to follow four structural trends: striking NE–SW, ENE–WSW, NNW–SSE and NW–SE.

(ix) The March–August 1983 flank eruption. The eruption was preceded by the escape of incandescent gas from the SE crater, and occurred along a fracture crossing the southern flank. A tiltmeter (L. Villari, pers. comm.) located at 2450 m, near the opening fracture, recorded a large ground deflation. Between 31 March and 1 April a double explosion crater opened near 2700 m above sea-level (asl) and phreatomagmatic ejections began which lasted a few days. A lava flow was emitted quietly but steadily from small vents; the most productive of which (at 2300 m asl) remained the only active vent after about two weeks of activity. The eruption was characterized by the extrusion of large lava fields, provoking the destruction of many buildings around the Rifugio Sapienza. The start of the eruption was preceded by a strong seismic crisis on 26 and 27 March (Fig. 4c) affecting the southwestern flank of the volcano, causing light damage to some villages located in the western flank and characterized by magnitudes in the range 2.5 to 3.0.

Patanè *et al.* (1984) noted that the epicentral distribution of the shallower events ($h < 3$ km) did not coincide with the location of the eruptive fractures, whose opening should have required a very small release of seismic energy. Deeper earthquakes ($h > 4$ km) trended ENE–WSW and NNW–SSE according to the stress field model of Lo Giudice *et al.* (1982). Intense seismic activity also occurred on 28–30 March, and other swarms were recorded on 13–15 May, and 3–5 June, with magnitudes between 2.5 and 3.2. Finally earthquakes with magnitude between 2.5 and 3.0 occurred during the final phase of the eruption, also affecting the eastern flank of the volcano. The swarm occurring on 3 and 4 June was due to deep events ($10 < h < 40$ km), with epicentral distribution strictly trending NNW–SSE. This swarm was preceded and followed by shallow earthquakes (Glot *et al.* 1984). An ENE alignment of epicentres was observed before the swarm, with events concentrated on the western side of the volcano at depths of less than 6 km. After the deep swarm, the activity seemed to migrate to the upper eastern side of the volcano and also affected the ENE–WSW structural trend.

(x) The April–October 1984 flank eruption. During the night of 27–28 April 1984, an eruptive fracture orientated approximatively NW–SE opened inside the SE crater, near its NE margin. Lava flows poured from the SE crater over the following 172 days, entering the Valle del Bove. The eruption was accompanied by intense, but not continuous, explosive activity involving the summit craters. Intense long-duration seismic activity, concentrated within the entire Etnean area, preceded, accompanied and followed the eruption (Fig. 4c).

Several earthquakes occurred between 10–23 April. Two isolated earthquakes occurred on 15 April and on 19 June causing property damage and affecting the western flank.

Remarkable seismic activity was recorded during the final phases of the eruption. First, a swarm of moderate ($2.5 < M < 3.0$), deep ($12 < h < 23$ km) events occurred from 25–30 September mostly concentrated on the southwestern side of the volcano. A sudden increase (Gresta *et al.* 1987) of the seismic activity due to shallow and weak ($2 < M < 2.5$) events was observed in the afternoon of October 16, mainly involving the northeastern flank of the volcano, while on 17 October a large number of shallow and weak ($2 < M < 2.5$) events affected the eastern and the southern flanks. Three strong earthquakes that provoked significant damage occurred respectively at Piano Provenzana on October 18 (Int. = VI–VII), at Zafferana on 19 October (Int. = VII) and at Milo on 25 October (Int. = VIII). It is important to stress that, as for the 1983 eruption, the aforementioned swarm of relatively deep events ($12 < h < 23$ km) occurred on the southwestern side of the volcano, and was probably linked to the NNE–SSW structural trend, which acted as a conduct for rising magma (Cristofolini *et al.* 1985).

(xi) The March–July 1985 flank eruption. Weak strombolian activity started on 8–9 March 1985 at the SE crater. Lava began to flow from the southeastern crater on the morning of 10 March and advanced eastward (toward the Valle del Bove), stopping some hours later. On 11 March fissures opened on the upper south flank at an elevation between 2300 and 2620 m. On 12 March an exclusively effusive vent opened at lower elevations (2600; 2510 and 2490 m, respectively). The lava flows that originated from these vents moved mainly towards the south and southwest, giving rise to numerous individual lobes, which roughly covered the 1983 flow field. The eruption lasted for 125 days, ending on 13 July 1985.

No increase of seismic activity occurred at the time of the eruption (Fig. 4c,d) and earthquakes recorded both before and during the eruption affected the eastern and the southern flanks, and were characterized usually by low magnitudes (1.5 < M < 2.5), and a gradual increase with time of the energy released. Two stronger events occurred on March 1, affecting the S. Venerina and Zafferana area (Int. = V–VI), and on March 23 affecting the southern flank (Int. = V–VI). The lack of significant seismic activity might be explained in terms of magma intruding a previously opened fissure system.

(xii) The December 1985 flank eruption. After Strombolian activity started at the SE crater on 19 December, an eruptive fissure opened on 25 December affecting the western boundary of the Valle del Bove, pouring out a lava flow which stopped to the NE of Mt Centenary. Eruptive activity from the fissures stopped early the following morning. A second phase occurred on 28 December with activity resuming from the same place. This phase was accompanied by weak Strombolian activity and stopped on 31 December. Appreciable seismic activity occurred during the months immediately preceding the eruption. A swarm of events with magnitudes between 2.5 and 3.5 occurred on 27 October, with a large number of earthquakes recorded also in November and at the beginning of January. The start of the eruption was accompanied by a strong seismic swarm, due to the fracture opening that occurred on the night of 25 December, consisting of earthquakes with magnitudes between 2.0 and 3.5 affecting the eastern flank and signalling another high energy release (Fig. 4d). The strongest event (Int. = VIII) located at Piano Provenzana, destroyed a hotel with one casualty. Another seismic swarm, with similar characteristics occurred on 26 December and the activity decreased gradually over the following days.

(xiii) The September 1986 subterminal activity. The eruption started at the NE crater on 13 September 1986 and was preceded by intense Strombolian activity at the Bocca Nuova which produced a 40 m high scoria cone within the crater. During the effusive phase, Strombolian activity varied in strength but generally remained at a low level. The eruption ended on 24 September 1986. An appreciable number of earthquakes giving a moderate energy release (Fig. 4d) characterized the months preceding the eruption, with magnitudes between 2.5 and 3.2, affecting the eastern and the western flanks. The start of the eruption was accompanied only by explosion-quakes probably related to the beginning of increased northeast crater activity. During the night of 22 September, three isolated seismic shocks with a maximum magnitude of 2.9, were recorded on the NW flank.

(xiv) The October 1986 flank eruption. This eruption occurred on 30 October along a 2 km long fissure system, orientated ENE, which stretched from the base of the Chasm to the west flank of the Valle del Bove. About eight hours after the start of the eruption, two boccas fed two main flows moving SE. During the early afternoon the fissure system continued to propagate downhill, creating a new eruptive fissure. On 31 October a lava flow was extruded from the SE crater during a single day. The effusive activity stopped at the two earlier boccas on 3 February 1987 and 1 March 1987. A seismic swarm (Fig. 4d) occurred during the night of 2 October (30 deep events with a maximum magnitude of 3.3) affecting the lower NW flank. A second swarm of deep events started during the morning of 5 October, again on the western flank, while another swarm of about 10 shocks occurred on the morning of 7

October. In both cases the magnitude maximum was 3.2. During the following days, both shallow and deep earthquakes occurred, mainly on the western flank of the volcano. On 29 October an intense but brief seismic swarm (40 shocks) preceded the eruption with magnitudes between 2.5 and 3.0. Only a few weak earthquakes were recorded during November 1986 and February 1987.

(xv) The September–October 1989 flank eruption. The eruption began on 10 September with explosions at the SE crater and at the Chasm. Thirteen strong eruptive episodes occurred between 11 and 26 September, including vigorous explosive activity from the SE crater with lava fountains reaching heights of > 1 km. On 24 and 25 September two eruptive fractures occurred along the SE and NE flanks of the SE crater, respectively. Two boccas opened on 27 September feeding lava flows which affected the eastern flank. From 27 September to 3 October a dry fracture opened which crossed the western wall of the Valle del Bove reaching the road Zafferana–Rifugio Sapienza at 1510 m asl. The eruption had a duration of 29 days, ending on 9 October. Seismic activity with magnitude between 2.0 and 3.2 occurred during the months preceding the eruption (Fig. 4e) mainly affecting the summit area. An increase in the seismic activity occurred at the end of August and on 3 September with magnitudes between 2.5 and 3.2. A swarm with a large number of earthquakes with magnitude ranging mainly from 2.5 to 3.5 occurred between 23 September and 3 October, while a sudden increase in the number of earthquakes on 2 October characterised the opening of a system of fractures with the strongest event reaching magnitude 3.7. A recent study (Ferrucci *et al.* 1993) shows that the epicentres of the earthquakes recorded by a temporary seismic array located in the area affected by the fracture shifted towards southeast. In November and December seismic activity returned to low levels.

(xvi) The January–February 1990 subterminal activity. A number of eruptive episodes occurred at the SE crater on 4–5, 11–12, 14–15 January and 1–2 February, characterized by intense explosive and Strombolian activity, lava fountaining and lava flow formations, affecting the high eastern flank of the volcano. No unusual seismic events were associated with the activity (Fig. 4e), with the exception of a few earthquakes at the end of January with magnitudes of around 2.5, mainly affecting the western flank. Significant increases of volcanic tremor were, moreover, recorded during the most intense phases of Strombolian activity (Montalto *et al.* 1992).

(xvii) The December 1991–March 1993 flank eruption. The eruption started on 14 December 1991 along a fracture on the northern side of the SE crater, also affected by explosive activity. At the same time, another eruptive fracture propagated from the southeastern base of the SE crater, characterized by strong explosive activity and the production of two small lava flows. The same fracture continued to propagate downhill and on 15 December two boccas opened on the western wall of the Valle del Bove, The lava flows crossed the Valle del Bove, reaching the Piano dell'Acqua which is located 1 km north of the town of Zafferana. This was the longest eruption during the period considered in this paper, lasting for 473 days. Four swarms preceded the eruption, on the 14, 15, 19, and 21–22 December located on the upper northern and western flanks, and apparently aligned in an ENE–WSW direction. During the period accompaning the opening of the eruptive fractures, isolated

earthquakes were recorded (19–20 December and 23 December–1 January 1992.

The most important cluster of events (more than 200 shocks), which produced the highest energy releases (Fig. 4f), occurred on 14 and 15 December, and an earthquake with magnitude 4.6 was felt at Catania and its surroundings. The focal mechanisms of some earthquakes computed (Patanè et al. 1996) suggest that three different phases occurred: the first and third phases characterized by distensive patterns, and the second by a compressive phase.

Main phases of seismic activity not correlated with eruptive events

Apart from the seismic activity occurring before, during or soon after eruptions, and treated in the previous section, some relevant episodes of seismic energy release have occurred over the period of interest which are not apparently associated with variations in the volcanic activity. The first such episode occurred in June 1981 (see Fig. 4b) and affected the lower western flank, reaching a maximum magnitude of 3.9 (22 June, 09:36 GMT). Another episode was recorded from 7–10 May 1986 (Fig. 4d), characterized by several tens of earthquakes of moderate magnitude (between 2.5 and 3.0) that occurred mainly on the eastern flank at depths between 5 and 15 km. Finally, a M = 4.1 earthquake occurred on 29 January 1989 (07:30 GMT) located on the lower eastern flank at a depth of about 2.6 km (Patanè et al. 1994), causing structural damage to the motorway connecting the cities of Catania and Messina.

Magnitude–frequency distribution

One of the fundamental statistical relationships applied to seismology is the magnitude–frequency distribution of earthquakes. Many empirical approaches have been proposed (Gutenberg & Richter 1956; Shilien & Toksoz 1970; Utsu 1971; Purcaru 1975). We studied the behaviour of the b coefficient of the Gutenberg–Richter (1956) relationship:

$$\text{Log } n(M) = a - bM$$

in the Etnean area during 1977–1991. The parameter b provides information about the distribution of magnitude within a sequence or a swarm. Its value decreases when the ratio between earthquakes with 'high' magnitude and earthquakes with 'low' magnitude increases. The significance of the coefficient b is underlined by many authors. Scholz (1968) assumes a significant dependence of b on rock stress, while recent studies have permitted the correlation of b and seismogenetic fault geometry (Aki 1965; Huang & Turcotte 1988; Hirata 1989).

The magnitude–frequency distribution computed for all the earthquake sequences at Etna shows a satisfactory fit for the magnitude range between 2.6 and 3.9 (Fig. 5a). The same result was obtained calculating the b value independently for the eastern (Fig. 5b) and the western flank (Fig. 5c). As a further approach, we investigated the variation of b over the whole period studied. Previous studies performed in the Etnean area to appreciate changes on the b values immediately before an eruption, gave information on the stress-field characteristics near eruption sites (Gresta & Pataneé 1983a, b). In the present study we aimed to evaluate the

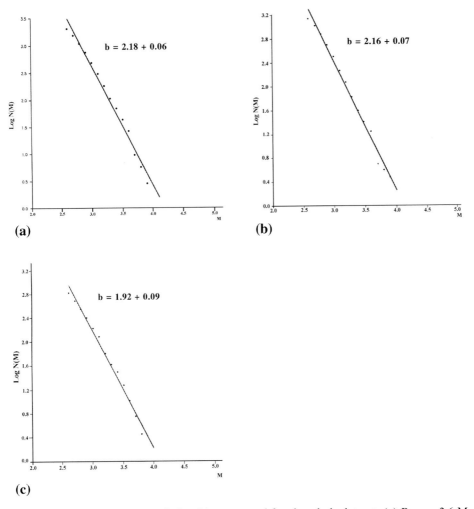

Fig. 5. Magnitude–frequency relationship computed for the whole dataset. (**a**) Range: 2.6 M 3.9; number of events: 5352; (**b**) range: 2.6 M 3.9. number of events: 3804; (**c**) range: 2.6 M 3.8; (**c**) number of events: 1547.

trend of b during a period which includes several eruptions. We estimated the b value, considering shifting 'windows' of 100 events with steps of 50 events. The next stage was to independently evaluate the b value for 'shallow' earthquakes ($0 < h < 7$ km) and for 'deeper' ones ($h \geq 7$ km) (Fig. 6a). The reference depth of 7 km was chosen according to the results of a study on focal earthquake mechanisms for the Etnean area carried out by Scarpa et al. (1983). For this approach we considered only those earthquakes with hypocentral errors less than 3 km. The b value computed for earthquakes with depths < 7 km shows a marked fluctuation and a steady decrease after 1988 with values between 0.7 and 1.3, whereas the b value for earthquakes with depth ≥ 7 km is almost stable around the average value (c. 1.7). Finally, we investigated the behaviour of b for the eastern and the western flanks separately (Fig. 6b). Again, we considered only earthquakes with epicentral

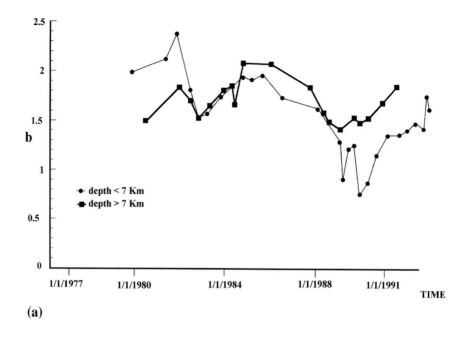

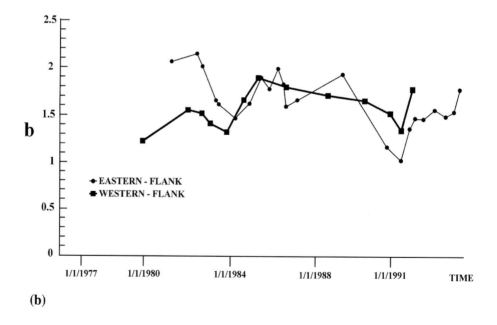

Fig. 6. Temporal evolution of b value computed for (**a**) different ranges of depth and (**b**) for the two flanks.

errors less than 3 km, and discovered that eastern flank b values show higher fluctuations than b values associated with the western flank.

Discussion and conclusions

It is well known that seismic energy release during the beginning eruptive and declining stages of volcanic activity provides information about the mechanisms and evolution of volcanic eruptions (Yokoyama 1988). In this respect possible relationships between earthquakes and eruptions at Etna between 1977 and 1991 are summarized below.

1. Summit eruptions are not immediately preceded by appreciable energy releases. Hill (1977) justifies this behaviour with a model whereby the migration of magma inside the volcano is associated with fracturing processes causing a gradual release of energy over time. Within such a hypothesis magma rises mainly aseismically through the main conduits and fractures, or by fracturing the enclosing rocks and producing microearthquakes. Nevertheless, we noticed that appreciable energy releases occur some months before this kind of eruption, suggesting that before eruption magma resides in small magmatic batches within a shallow stress field. The energy release plots of Figs 4 (a,b,c,d,e) suggest that the occurrence of such magma transfer is not completely aseismic.
2. The duration of summit eruptions generally became longer over the period of interest with time. Seismic energy releases before summit eruptions increased with time, with particular reference to the eruptions occurring at the SE crater. This fact is confirmed by the appreciable rupture episode that occurred on the low eastern flank on 29 January 1989. Moreover, eruptions occurring at the SE crater were characterized by a longer duration than those at the NE crater. It is remarkable that, after the NE crater eruptions occurring in 1977, which were characterized by a very low energy release, the SE crater was constructed, accompanied by strong energy release on 29 April 1978. It is suggested that a change, probably induced by a variation of the Etnean stress-field, occurred within the plumbing system of the volcano at this time, since the following eruptions predominantly affected the SE crater.
3. The onset of flank eruptions is accompanied by significant energy release due to the opening of the eruptive fractures, with the exception of the March–July 1985 eruption that took place on the same fracture system which fed the 1983 eruption.
4. Earthquakes of appreciable magnitude characterized the ends of the eruptions of 1981, 1983, 1984 and 1986, according to Nercessian *et al.* (1991). This behaviour can be explained in terms of the edifice redistribution of stresses and pressures in the heterogeneous mass.
5. Two significant seismic energy releases, which occurred on June 1981 and May 1986 are not apparently linked to eruptions.
6. The b values computed for depths <7 km are characterized by a significant decrease during 1988 with a minimum during 1989. In the following period the b value experienced a general increase. The b value for depths ≥ 7 km is characterized by a regular behaviour with fluctuations within error bars

(average = 1.7), possibly suggesting that the deeper stress-field is not heterogeneously distributed within the etnean basement. Furthermore, the temporal evolution of b, evaluated independently for the eastern and the western flanks, is in agreement with the structural models proposed for the volcanic edifice. The models point out the seaward displacement of the eastern flank of the volcano pile, unbuttressed on a clay-rich layer, in contrast to the remainder of the edifice, which is topographically buttressed and which rests on a more competent substratum. It implies, in terms of seismicity, a larger number of earthquakes with a lower magnitude affecting the eastern flank respective to the western one. Hence, the b values, relative to the eastern flank, are generally higher and fluctuate in a wide range and respect to the western one. These results confirm the findings of a recent analysis (Montalto et al. 1996) supporting that the eastern flank is influenced by more intense geodynamical activity, probably related to both tectonics and flank instability processes.

References

AKI, K. 1965. Maximum likelihood estimate of b in the formula Log $N = a - bM$, and its confidence limits. *Bulletin of the Earth Resources Institute, Tokio University*, **43**, 237–239.

AZZARO, R. & NERI, M. 1992. *The 1971–1991 eruptive activity of Mt. Etna (First steps to relational data-base realization)*. CNR, IIV, Open File Report, 3/92.

CRISTOFOLINI, R., GRESTA, S., IMPOSA, S., PATANÈ, G. 1985. Feeding mechanisms of eruptive activity at Mt. Etna based on seismological and petrological data. *In:* KING, C. Y. & SCARPA, R. (eds) *Modelling of Volcanic Process*. Friedr. Vieweg & Sohn, Braunschweig/Wiesbaden. 73–93.

FERRUCCI, F., RASÀ, R., GAUDIOSI, G., AZZARO, R. & IMPOSA, S. 1993. Mt. Etna: a model for the 1989 eruption. *Journal of Volcanic and Geothermal Research*, **56**, 35–56.

GASPERINI, P., GRESTA, S. & MULARGIA, F. 1990. Statistical analysis of seismic and eruptive activities at Mt. Etna during 1978–1987. *Journal of Volcanic and Geothermal Research*, **40**, 317–325.

GLOT, J. P., GRESTA, S., PATANÈ, G. & POUPINET, G. 1984. Earthquake activity during the 1983 Etna eruption, *Bulletin of Volcanology*, **47**, 953–963

GRESTA, S. & PATANÈ, G. 1983a. Variations of b values before the Etnean eruption of March 1981. *Pure and Applied Geophysics*, **121**, 287–295.

—— & —— 1983a. Changes in b values before the Etnean eruption of March–August 1983. *Pure and Applied Geophysics*, **121**, 903–912.

——, GLOT J. P., PATANÈ G., POUPINET G. & MENZA, S. 1987. The October 1984 seismic crisis at Mt. Etna. Part 1: space-time evolution of the events. *Annals Geophysicae*, **5B**, 671–680.

GUTENBERG, B. & RICHTER, C. F. 1956. Magnitude and energy of earthquakes, *Ann. Geof.* **9**(1), 1–15.

LO GIUDICE, E., PATANÈ, G., RASÀ, R. & ROMANO, R. 1982. The structural framework of Mount Etna. Memorie Società Geologica Italiana, **23**, 125–158.

HILL, D. P. 1977. A model for earthquake swarms. *Journal of Geophysical Research*, **82**, 1347–1352.

HIRATA, T., 1989, A correlation between the 'b' value and the fractal dimensions of earthquakes. *Journal of Geophysical Research*, **94**(B6), 7507–7514.

HUANG, J. & TURCOTTE, D. L. 1988. Fractal distributions of stress and strength and variations of b-value. *Earth and Planetary Science Letters*, **91**, 223–230.

MONTALTO, A., DISTEFANO, G. & PATANÈ, G. 1992. Seismic patterns and fluid-dynamic features preceding and accompanying the January 15, 1990 eruptive paroxism on Mt. Etna (Italy). *Journal of Volcanology and Geothermal Research*. **51**, 133–143.

——, VINCIGUERRA, S., MENZA, S. & PATANÈ, G. 1996. Recent seismicity of Mount Etna : implications for flank instability. *In:* McGUIRE, W. J., JONES, A. P. & NEUBERG, J. (eds) *Volcano Instability on the Earth and other Planets.* Geological Society, London, Special Publications, **110**, 169–177.

NERCESSIAN, A., HIRN, A. & SAPIN, M. 1991. A correlation between earthquakes and eruptive phases at Mt. Etna: an example and past occurrences. *Geophysics Journal International*, **105**, 131–138.

PATANÈ, G., GRESTA, S. & IMPOSA S. 1984. Seismic activity preceding the 1983 eruption of Mt. Etna, *Bulletin of Volcanology.* **47**, 941–952.

——, MONTALTO, A., IMPOSA, S. & MENZA, S. 1994. The role of regional tectonics, magma pressure and gravitational spreading in earthquakes of the eastern sector of Mt. Etna volcano. *Journal of Volcanology and Geothermal Research*, **61**, 253–266.

——, ——, VINCIGUERRA, S. & TANGUY J. C. 1996. A model of the 1991–1993 eruption onset of Etna (Italy). *Physics of the Earth and Planetary Interviews*, **97**, 231–245.

PURCARU, G. 1975. A new magnitude–frequency relation for earthquakes and a classification of relation types. *Geophysical Journal of the Royal Astronomical Society*, **42**, 61–79.

RICHTER, C. F. 1958. *Elementary Seismology.* Freeman and Company, San Francisco.

SCARPA, R., PATANÈ, G. & LOMBARDO, G. 1983. Space-time evolution of seismic activity at Mount Etna during 1974–1982. *Annales Geophysicae*, **1**, 451–462.

SCHICK, R., LOMBARDO, G. & PATANÈ, G. 1982. Volcanic tremors and shocks associated with eruptions at Etna (Sicily), September 1980. *Journal of Volcanology and Geothermal Research*, **14**, 261–279.

SCHOLZ, C. H. 1968. The frequency–magnitude relation of microfracturing in rock and its relation to earthquakes. *Bulletin of the Seismological Society of America*, **58**, 399–415.

SEAN, Bulletin. 1978, 1979, 1981, 1984, 1985, 1986, 1989, 1991. Smithsonian Institute, Washington.

SHARP, A. D. L., LOMBARDO, G. & DAVIS, P. M. 1981. Correlation between eruptions of Mount Etna, Sicily, and regional earthquakes as seen in historical records from AD 1582. *Geophysical Journal of the Royal Astronautical Society*, **65**, 507–523.

SHILIEN, S. & TOKSOZ, M. N. 1970. Frequency–magnitude statistics of earthquakes occurrences. *Earthquake notes*, **41**, 5–18.

TANGUY, J. C. & PATANÈ, G. 1984. Activity of Mount Etna, 1977–1983: volcanic phenomena and accompanying seismic tremor. *Bulletin of Volcanology*, **47**, 965–976.

UTSU, T. 1971. Aftershocks and earthquakes statistics, III. *Journal of the Faculty of Science of Hokkaido University, Series VII, Geophysics*, **3**(5), 379–441.

YOKOYAMA, I. 1988. Seismic energy releases from volcanoes. *Bulletin of Volcanology*, **50**, 1–13.

Modelling the impact of Icelandic volcanic eruptions upon the prehistoric societies and environment of northern and western Britain

JOHN GRATTAN,[1] DAVID GILBERTSON[2] & DANIEL CHARMAN[3]

[1] *The University of Wales, Institute of Geography & Earth Sciences, Llandinam Building, Aberystwyth SY23 3DB, UK*
[2] *The Nene Centre for Research, Nene University College, Northampton NN2 7AH, UK*
[3] *The University of Plymouth, Department of Geographical Sciences, Drake Circus, Plymouth, PL4 8AA, UK*

Abstract: Many studies now address the impact of Icelandic volcanic eruptions upon the societies and environment of Britain and Ireland. It has become apparent that the assumptions of the magnitude of volcanic impact inherent in these studies are open to question. The scale of climatic change following many eruptions has been observed to be ephemeral or non-existent, whilst the impact of toxic volatile gases and aerosols is dependent, to a large degree, on the vulnerability of the receptor rather than the scale of deposition. The scale and extent of volcanic climatic-forcing mechanisms are examined and the popular perception of the effectiveness of these is challenged. In particular, the concepts of harsh volcanic winters and cool volcanic summers is examined via case studies of the impacts of the three largest eruptions of recent and historical times, i.e. those of the Laki fissure, Tambora and Mount Pinatubo. Alternative volcanic forcing mechanisms for environmental change are considered, principally those concerned with the transport and deposition of toxic volatile material. The need for caution in the adoption of these models is also urged. It is suggested that workers in this field need to adopt a clearer theoretical framework when contemplating the association of palaeoenvironmental data and volcanic events. Such a framework is proposed.

Many studies have suggested that volcanic eruptions may be capable of affecting the environment and societies of northern Europe and, in particular, of northern Britain and Ireland (Baillie & Munro 1988; Burgess 1989; Blackford *et al.* 1992; Grattan & Charman 1994). This research has been driven by the discovery of tephra layers from different volcanic provinces in lake, mire and peat ecosystems, which are traditionally exploited in palaeoenvironmental studies. In Britain, this approach was pioneered by Dugmore (1989), and an extensive tephrochronology is now being developed which covers much of the British Isles and northern Europe (Pilcher & Hall 1992; Lotter & Birks 1993; Bogaard *et al.* 1994; Hall *et al.* 1994; Raynal *et al.* 1999).

It is perhaps inevitable that palaeoenvironmental and archaeological studies, focused by tephra isochrones (Buckland *et al.* 1981), have sought to identify or isolate a volcanic signal in the data recovered. Where this is the case, caution must be urged if any data which suggest environmental change, noted in an intensively

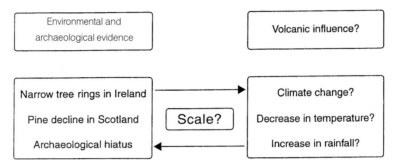

Fig. 1. Deterministic linkage of environmental phenomena and volcanic eruptions.

studied sample, are not to be casually ascribed to the environmental forcing of a distant, but temporally convenient, volcanic event. On a similar theme, Renfrew (1979) sounded a note of caution:

> ...it is necessary to recognise and discount the common tendency among archaeologists and historians to assume a causal link between the distant and widely separated events of which they may have knowledge. An eruption here, a destruction there, a plague somewhere else – are all to easily linked in hasty surmise.

In the context of palaeoenvironmental studies which seek to establish the role of volcanic eruption in environmental change in Europe, a further caveat can be proposed:

> Care must be taken to avoid making the assumption that the external forcing mechanism identified was of sufficient magnitude to bring about the environmental change observed.

The dangers inherent in such an approach are illustrated in Fig. 1; environmental and archaeological evidence is gathered, a convenient volcanic eruption identified and a volcanic influence of a sufficient scale to account for the fluctuation in the data is assumed. It is clear that, in this conceptual framework, all issues of scale and magnitude are ignored or 'glossed over' and it is with this issue that this paper is primarily concerned.

Volcanic eruptions, archaeology and environment in northwest Europe: a review

A brief review of typical studies, which propose a volcanic mechanism for environmental change, is necessary before the mechanisms, which may be responsible, are reviewed critically. Burgess (1989) discussed an apparent hiatus in settlement history in northern Scotland in the late Bronze Age and suggested a possible volcanic mechanism for this event, the Hekla 3 eruption, a theory later reiterated by Barber in a BBC television programme entitled *A Time of Darkness*. This phenomenon is particularly well illustrated in the straths and glens of northeast Scotland; along the Strath of Kildonan, up to 2000 hut circles appear to have been abandoned in a very short period of time. A catastrophic forcing mechanism for this phenomenon has long been favoured (Burgess 1985) and the temporally convenient environmental stress identified in the Irish dendrochronology (discussed below) has led to speculation that the Hekla 3 eruption was responsible (Burgess 1989).

Environmental stress has been identified in several studies. In the Irish–Bog-oak chronology, Baillie and others (Baillie 1988, 1989*a, b*, Baillie & Munro 1988) have identified several episodes of distinct environmental stress, which appears to persist for decades, in response to volcanic eruptions which are dated by ice-core acidity records (Hammer 1977, 1984; Hammer *et al.* 1980, 1981, 1987; Herron, 1982). Blackford *et al.* (1992) and Edwards *et al.* (1994) have identified a clear decline in the pollen of *Pinus sylvestris* in northern Scotland which appears to occur in tandem with the deposition of tephra from the Hekla 4 eruption. In contrast, however, in the Strath of Kildonan, Charman *et al.* (1995) believed that they had noted an increase in Pinus pollen with the same event, while in Ireland, Hall *et al.* (1994) could not identify any response in the Pine record to the Hekla 4 event. Charman *et al.* (1995) also noted a decline in the pollen of *Corylus* in tandem with an unidentified tephra, and also the occurrence of other tephra-falls, which generated no apparent response in the pollen record. In Shetland, Bennett *et al.*, (1992) also noted a decrease in arboreal pollen with an apparent correlation to a tephra-fall, though they carefully observed that this decline also occurred in tandem with a charcoal influx and may therefore be anthropogenic rather than volcanogenic in origin. Tephra-fall also appears to have had an impact on Lake Ecosystems. Edwards *et al.* (1994) noted a severe, but short-lived decline in the abundance of Chironomidae in Skælingisvatn, a lake in The Faeroes. In southern Germany, Lotter & Birks (1993) observed a similar short-lived decline in diatom abundance following the deposition of the Laacher see tephra.

An alternative approach to the investigation of volcanic impacts upon the European environment focuses upon the direct observation of volcanogenic environmental change via the study of historical documents. Thórarinsson (1981) described the impact of gases and aerosols from the Laki fissure eruption in Scandinavia. Grattan and others (Grattan 1994; Grattan & Charman 1994, Grattan & Gilbertson 1994; Grattan & Pyatt 1994; Grattan & Brayshay 1995; Brayshay & Grattan 1999) have researched British documentary records of this event in great detail, and illustrated the potent social and environmental influence which the gases and aerosols from this event were able to wield across Europe, causing damage to vegetation and engendering unease in people from Aberdeen to Naples. Camuffo & Enzi (1995), adopting a similar approach to the Italian documentary record, have also discussed the severe distal environmental impact of volcanic gases emitted into the troposphere by Italian volcanoes.

The studies briefly reviewed above suggest that two principal mechanisms exist by which volcanic eruptions may bring about environmental change in distal environments. The first is via climatic change and the second via the deposition of toxic volatile material. However, the work reviewed above also illustrates a growing problem. Researchers are identifying tephras and often proposing volcano related environmental change. If climatic mechanisms are to be invoked it would seem reasonable that a degree of consistency should be apparent between the environmental impact of any single volcanic eruption. Equally, if acid impacts are to be invoked, the sensitivity of plant species or ecosystems to this mechanism should be considered before attempting to construct a regional or national model of volcanic impacts.

As more tephra–environmental studies are conducted, it has become apparent that there appears to be a growing disparity between the results presented and the environmental change identified, whereby different studies do not appear to be consistent at the regional scale. Caseldine *et al.* (1998) elegantly illustrated this problem. The same tephra, studied in three cores taken in close proximity, presented contradictory evidence of environmental change. These problems need to be resolved, as it is clear that the mechanisms by which volcanic eruptions have an impact upon the environment are not understood, nor is the scale at which these mechanisms operate. Buckland *et al.* (1997) have also sounded a similar note of caution in a recently published paper.

The principal volcano–climate relationships are now examined in detail in an attempt to establish whether such mechanisms lend themselves to the development of models of hemispheric environmental change (Baillie 1989*b*). The role of volcanic eruptions in creating environmentally extreme weather, volcano winters (Rampino 1988) and volcano summers (Stommel & Stommel 1979) is examined.

Volcano weather

Winter temperatures in Britain: the Edinburgh temperature record

Mossman's (1896) collation of temperature records for Edinburgh allows a reconstruction of temperature trends. It is obvious that the winter of 1783–84, based on the means of January and February, was not exceptional. In the period 1764–1830 there were seven other winters, which were as cold, or colder, than that associated with the eruption of Laki. The conclusion that must be drawn is that the winter temperatures in Edinburgh associated with the eruptions of the Laki fissure and Tambora were not exceptional. The year concerned can be seen to be part of a temperature series in which extremes of cold and warmth fail to correlate with volcanic events. Volcanic eruptions alone cannot be held solely responsible for causing cold winters in Scotland in the late eighteenth and early nineteenth centuries.

Winter temperatures in Britain: the central England temperature record

In England, as in Scotland, analysis of temperature records indicates the occurrence of harsh winters in years, which are **not** associated with recorded volcanic eruptions (Manley 1974). Mean temperatures reveal that the winter following the eruption of Laki was the fourth most severe between 1733–1833, but that it was also part of a cooling trend in winter air temperatures. The years following Tambora cannot be considered exceptional. Any suggested correlation of **one** volcanic eruption with one of 14 winters between 1733–1833 which falls beyond one standard deviation below the mean must be considered of dubious significance, and equal caution must be exercised in the interpretation and correlation of palaeoenvironmental data with volcanic eruptions.

Winter temperatures in Europe: the Mount Pinatubo eruption

The climatic impacts of the Mount Pinatubo eruption are the most intensively studied in history, the impact of this event on European winter temperatures presents a further challenge to models which assume a severe winter to account for archaeological or environmental stress. Rather than a fall in temperatures, warming was in fact observed, the result of a disruption to the zonal wind (Graf *et al.* 1992, 1993). Robock & Mao (1992, 1995), in detailed studies of temperature records, have now observed an increase in winter surface-air temperatures in Europe, northern Asia and North America in the first winter following a low-latitude eruption and the second winter following a northern hemisphere high-latitude eruption. It is clear that neither of the most notorious eruptions of recent historical times nor the most intensively studied recent volcanic eruption, can support the contention that severe deterioration in winter conditions will follow major volcanic eruptions. Indeed, the most recent studies suggest that warming may occur. Therefore, caution must be used when speculating upon the climatic effect of Icelandic volcanic eruptions such as the Hekla 3 and Hekla 4 events.

Summer temperatures in Europe: The Laki fissure eruption

The summer of 1783 was the hottest in Manley's (1974) study of central England temperatures and high temperatures were reported across Europe. This phenomenon is discussed in detail in Grattan & Sadler (1999).

Summer temperatures in Europe: The Tambora eruption

It has been suggested that the presence of a stratospheric aerosol will disrupt the thermal gradient between the Equator and the poles, and between these and the land (Lamb 1970, 1977, 1988; Handler 1989). In the North Atlantic, this disruption is expected to displace the North Atlantic low-pressure minimum southwards from its summer position in the vicinity of Iceland towards the European mainland. This pattern of displacement was observed in the summer of 1816 (Kelly *et al.* 1984; Lamb 1992), ascribed to the influence of the Indonesian volcano Tambora which erupted in 1815. In 1816 many areas of mainland Europe experienced a cold, wet summer and famine was endemic in many regions (Wood 1965; Post 1977; Kington 1992), and in North America 1816 has entered folklore as 'the year without a summer' (Stommel & Stommel 1979, 1983; Harington 1992). Contemporary British documents reinforce our perception of 1816 as a year of widespread climatic and environmental stress:

> Melancholy accounts have been received from all parts of the continent of the unusual wetness of the season; property in consequence swept away by the inundation and irretrievable injuries done to the vine yards and corn crops. In several provinces of Holland, the rich grasslands are all under water, and scarcity and high prices are naturally apprehended and dreaded. In France the interior of the country has suffered greatly from the floods and rains
> *The Norfolk Chronicle*, July 20, 1816.

Should the present wet weather continue, the corn will inevitably be laid, and the effects of such a calamity at such a time cannot be otherwise than ruinous to the farmers and even the people at large.
The Times, July 20, 1816.

It would be reasonable to assume from this material that the climatic deterioration in Europe was widespread and thus models are developed for similar cold, wet summers to account for palaeoenvironmental variability and manifestations of archaeological stress. However, to upset this convenient picture, northern Europe appeared to have experienced distinctly different conditions. In Denmark, Sóren Pedersen (in Newmann 1990) recorded:

> This year was again a lovely fertile year, not only on account of the quantity of fodder, but also because the kernels were so big. This was important for our country as this commodity was very much coveted in foreign lands ... where there has been a general crop failure.

Temperature data presented in Neumann (1990, 1992) suggests that 1816 was in fact warmer than average in many northern European cities. Both environmental conditions are reconcilable. The air-pressure maps of Kelly *et al.* (1984) and Lamb (1992) illustrate the southwards displacement of the 'Icelandic Low', while Kington's (1992) rainfall isohyets illustrate the consequence of this displacement; regions to the south of the low-pressure minimum experienced wetter and colder summer conditions, whilst regions to the north, including northern Scotland, experienced dryer than normal conditions and a milder summer.

Wetter and colder summer conditions cannot be assumed to account for changes in the palaeoenvironmental and archaeological records, and it is apparent that while some areas undoubtedly experienced harsher than normal conditions, checks and balances operate which bring milder conditions to other areas. One should also note that Kelly *et al.* (1984) observed similar circulation disruption in other years where no volcanic forcing could be suggested. It is particularly important to note that in 1816 northern Scotland experienced a better summer than normal and assumptions of Scottish settlement abandonment in the face of climatic deterioration must be challenged.

Global temperature reduction

Observations of temperature response to radiative forcing by volcanic aerosols also fail to suggest that an environmentally significant reduction in surface-air temperature will occur. Following the Mount Pinatubo eruption, a globally averaged reduction in temperatures was observed not to exceed 0.5°C, which fell well within the normal range of global temperature fluctuation (Luhr 1992; Minnis *et al.* 1993; Kerr 1994; McCormick *et al.* 1995)

Since Pinatubo had a minimal effect, the impact of Icelandic volcanic eruptions in terms of their ability to sufficiently depress hemispheric temperatures, and thus to bring about the environmental stress ascribed to them needs to be assessed. Estimates of total acid emission, based on the ice-core signal, suggest that both the Hekla 3 and Hekla 4 eruptions emitted *c*. 60 million tonnes (Hammer *et al.* 1980).

More detailed analyses of the Hekla 3 eruption are now available and indicate yields of: 1.58×10^{11} g Sulphur; 2.48×10^{11} g Chlorine; 4.84×10^5 tonnes H_2SO_4: 2.55×10^5 tonnes HCl (Devine et al. 1984); a total acid output of 7.39×10^5 tonnes (Palais & Sigurdsson 1989).

Devine et al. (1984) and Sigurdsson (1990) correlated sulphur output and surface temperature response for nine major eruptions, which had occurred since 1783, and established the existence of a relationship between the two variables expressed by the equation:

$$\text{Temperature decrease (°C)} = (5.89 \times 10^{-5}) \times \text{sulphur output (g)}^{0.308} \quad R = 0.92.$$

Taking as an estimate for the sulphur output of the Hekla 3 eruption as 1.58×10^{11} g (Devine et al. 1984), the predicted temperature response to the Hekla 3 eruption is of the order of 0.1°C. A temperature reduction of this scale is indistinguishable from normal background fluctuations and can only be regarded as unlikely to have brought about severe environmental degradation (or loss of settlement) in northern and western Britain (Hansen & Lacis 1990; Schneider 1994).

Volcano weather: synthesis

There is, therefore, little theoretical or empirical justification for the suggestion that any volcanic eruption of the recent Holocene, and certainly none of the Hekla eruptions, were of a magnitude which might have generated medium- or long-term severe climatic deterioration, or the volcanically induced equivalent of a 'nuclear winter' (Turco et al. 1982, 1983, 1984; Rampino 1988). It appears probable that neither the Hekla 3 nor the Hekla 4 eruption would have been able to generate a fall in surface temperature greater than a few tenths of a degree centigrade. When palaeoenvironmental data is considered the probable magnitude of volcanic climate forcing must be realistically assessed before this possibility is invoked as a forcing mechanism.

Volcanic gases and aerosols

The role of volcanic gases emitted in Icelandic volcanic eruptions has recently been the subject of much research and speculation (Edwards et al. 1994; Grattan 1994). The long-range transport of modern industrial pollution and its deposition in locally significant concentrations has been recorded in the literature (Scorer 1992), and there is no reason to believe that volcanic eruptions cannot be interpreted as a similar point source. Damage to vegetation across a large area of Europe in 1783 has been attributed to gases emitted in the Laki fissure eruption. Documentary descriptions of 'dry fogs' and severe acid impacts on vegetation have been detailed in several papers (Grattan & Charman 1994; Grattan & Gilbertson 1994; Grattan & Pyatt 1994). While acid aerosol impact is clearly a powerful mechanism, which may generate extreme environmental change, care must again be exercised in its application. The conversion of acid gases to aerosols will only occur under favourable conditions which may not occur everywhere; conversely, micrometeorological conditions may intensify the toxic impact of acid volatiles (Wisniewski 1982).

Local soil conditions and the physiology of particular plant species play a crucial role in conditioning environmental response to acid deposition. The crucial concept is that each ecosystem has a 'critical load', the concept assumes that:

> ... there is a damage threshold for the response of ecosystems to acidic deposition. This may vary between different ecosystem and species (pollution 'receptors'). The critical load for a particular receptor–pollutant combination is defined as the highest deposition load that the receptor can withstand without long term damage occurring.
> (Bull 1991, p. 30).

Critical loads in northern and western Britain

The impact of acid deposition will be severe, with the greatest possibility of long-term damage, in those regions which have a low critical load. Large areas of upland in northern Scotland and Ireland are sensitive to low levels of acid deposition, since they have a low critical threshold and have a limited capacity for recovery (Goreham 1987; Skiba et al. 1989; Bull, 1991; Smith et al. 1993). In Scotland, Skiba et al. (1989) demonstrated that peats with the highest acidity and lowest base saturation were those where deposited acidity was $>0.8 \text{ kg H}^+ \text{ ha}^{-1} \text{ year}^{-1}$, and Bull (1991) has mapped large areas of Scotland and Ireland as sensitive to the deposition of relatively minimal levels of acid, $<0.5 \text{ Keq H}^+ \text{ ha}^{-1} \text{ year}^{-1}$. It is important to recognize that those regions in Ireland and Scotland which are mapped as having a low critical load are also those which appear to have exhibited palaeoenvironmental or archaeological stress in response to volcanic eruptions in Iceland. There appears to be a distinct possibility that noxious and toxic acid volatiles, discharged in Icelandic volcanic eruptions, may have been deposited far downwind in quantities which were notably in excess of the critical load of these already acidified ecosystems. Approximately 1 tonne km^2 of tephra, associated with the Hekla 4 ash, was calculated to have fallen in parts of Northern Ireland by Pilcher & Hall (1992). If it is accepted that, at this distance from eruption source, the volume of adsorbed volatiles approached the mass of the tephra (Oskarsson 1980), then no less than 50 times the annual critical load for the Irish sediments may have been deposited in one very brief period of time. The ecological and human consequences are therefore likely to have altered in severity according to the location and pre-existing acidity status of different areas.

Modelling the impact of volcanic eruptions on distant societies and ecosystems

The discussions above illustrate the problems which have become apparent in studies which attempt to utilize volcanic eruptions to account for palaeoenvironmental or archaeological stress. Unless the specific conditions which render the receptor vulnerable to volcanic forcing are identified, and assessed, such associations must necessarily be of dubious value. It is assumed that there is a clear temporal relationship between the volcanic event and the environmental response. In addition, the effectiveness of any individual volcanic eruption as an agent of environmental change must be considered; in particular, a realistic assessment of the degree of climate change must be made.

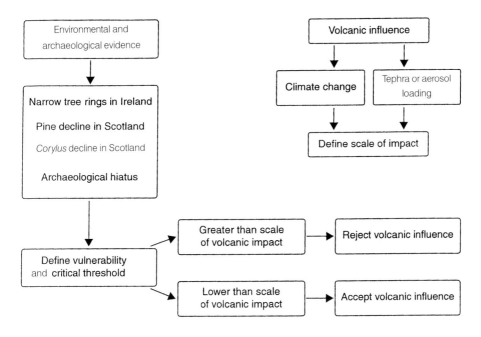

Fig. 2. A model by which to assess the relationship between volcanic activity and environmental response.

A simple flow diagram can be used to illustrate this approach (Fig. 2), in which three steps are involved:

1. the environmental and/or archaeological evidence for a volcanic influence must be established, and the areas in which the archaeology and flora or fauna may be sensitive to their environment, and the critical thresholds at which these operate should be defined;
2. the magnitude of the volcanic eruption and the scale of climate change and/or tephra–aerosol loading must be established;
3. if the critical threshold for the environmental/archaeological data is greater than the scale of volcanic impact then the volcanic influence should be rejected. If the critical threshold for the environmental archaeological data is less than the scale of volcanic impact then a volcanic influence may be considered as possible.

This approach is illustrated by two examples: A, the forcing of an environmental response in the Irish bog–oak chronology by the Hekla 4 eruption (Fig. 3); and B the response of Bronze Age settlement in northeast Scotland to the Hekla 3 eruption (Fig. 4).

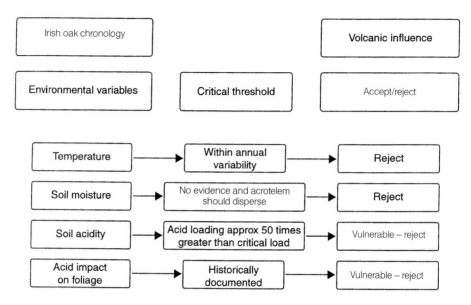

Fig. 3. Modelling the Hekla 4 eruption and Irish bog–oak response.

Case study A: the Irish bog–oak chronology and the Hekla 4 eruption

A detailed description of the environment enjoyed by the bog oaks can be found in Pilcher (1990), and the link between dendrochronological indicators of stress in Ireland and volcanic activity are well established. Four critical environmental variables have been identified for this example; surface-air temperature, soil moisture, soil acidity and direct impact of acids on foliage (Fig. 3); of course, it may be possible to consider other variables in the same way. The temperature response to an eruption of this magnitude is minimal and can be rejected (Palais & Sigurdsson 1989). There is no evidence for a long-term destabilization of atmospheric circulation patterns in response to similar and larger volcanic eruptions, and this possibility may also be rejected (Chester 1988; Mass & Portman 1989; Bradley & Jones 1992). Excess rainfall is unlikely to have severely affected the mire ecosystem, as mire hydrologists believe that the upper surface of an ombotrophic mire, the acrotelm, should effectively disperse excess water (Ingram 1978, 1987). In contrast, in terms of the base status, the mire sediments have a very low critical load, and thus the acid deposition theoretically associated with the tephra from the Hekla 4 eruption could have been far in excess of the buffering capacity of the ecosystem (Grattan & Charman 1994). Similarly, acid damage to foliage in the wake of historically documented volcanic eruptions has been recorded and this mechanism can be accepted as a possible cause of environmental degradation. By this process, a series of environmental variables were considered, their critical thresholds

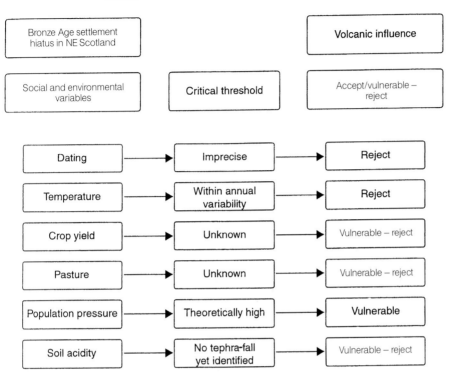

Fig. 4. Modelling the Hekla 3 eruption and settlement response in northeast Scotland.

established and a possibly effective volcanic forcing mechanism (acidification of sediments and foliage) proposed.

Case study B: Bronze Age settlement in northeast Scotland and the Hekla 3 eruption

No tephra from the Hekla 3 event has yet been positively identified in northeast Scotland and the association of the volcanic event with the imprecisely dated, and as yet unpublished evidence of a, synchronous abandonment of settlement are sufficient grounds to reject a volcanic forcing of settlement until further corroborative evidence is found. Modelled temperature response to the Hekla 3 event is within annual variability and can be rejected as an effective forcing mechanism (Fig. 4). The extent to which the agricultural and social pressures operating upon the societies of the region rendered them vulnerable to sudden environmental change is a subject of speculation (Barclay 1985; Burgess 1985) but in the absence of further evidence, must be rejected. As in Ireland, the sediments of the region can largely be classified as vulnerable to acid deposition; no tephra-fall has yet been established nor any palynological response identified and, as a result, a purely volcanic forcing mechanism appears doubtful. However, Baker *et al.* (1995) have identified growth irregularities in Scottish Speleothems and have speculated that this may be due to acid deposition following the Hekla 3 eruption. This does not link Hekla 3 to the putative settlement abandonment in northeast Scotland. Current data therefore suggests that while the environment of north-east Scotland is vulnerable to volcanic forcing, there are currently insufficient grounds to support the hypothesis that the Hekla 3 event was responsible for the late Bronze Age settlement crisis in the region.

Conclusions

This paper has attempted to introduce a degree of rigour into the use of volcanic forcing mechanisms to account for sudden environmental change in northwestern Europe and, in particular, in Britain and Ireland. The principal mechanisms by which volcanic eruptions may affect climate have been reviewed and a series of case studies presented which illustrate both the transient and often ephemeral nature of volcano related climate change, but also that such change has a strong regional bias. When volcano climate mechanisms are considered, the possibility of a climatic deterioration should not be considered automatically, since there is abundant evidence in the literature to suggest that climatic amelioration or improvement may also occur.

Where acid impacts – either of acids adsorbed on tephra shards or of acid aerosols – are considered, all available evidence suggests that such impacts are strictly local in occurrence, controlled by local micro-climatic factors, as well as the vulnerability of different receptors, soils or vegetation, to the acid loading.

A simple model has been proposed which encourages a critical assessment of the data, which relate environmental phenomena to volcanic events. This model not only allows such influences to be accepted/rejected, but also serves to point out those environmental variables which are vulnerable to volcanic influence and would reward further attention.

This work suggests that any attempt to construct regional, national or continental models which detail volcanic forcing of the environment must be based on a synthesis of many studies which present a palaeoenvironmental analysis of the impact of the same volcanic event, rather than by extrapolating outwards from the environmental response apparent in a single study.

References

BAILLIE, M. G. L. 1988. Irish oaks record volcanic dust veils drama. *Archaeology in Ireland*, **2**, 71–74.
——— 1989*a*. Do Irish bog oaks date the Shang Dynasty. *Current Archaeology*, **117**, 310–313.
——— 1989*b*. Hekla 3, how big was it? *Endeavour, New Series*, **13**, 78–81.
——— & MUNRO, M. A. R. 1988. Irish tree rings, Santorini and volcanic dust veils. *Nature*, **322**, 344–346.
BAKER, A., SMART, P. L., BARNES, W., EDWARDS, L. & FARRANT, A. 1995. The Hekla 3

eruption recorded in a Scottish Speloethem? *The Holocene*, **5**, 336–342.

BARCLAY, G. J. 1985. Excavations at Upper Suisgill, Sutherland. *Proceedings of the Society of Antiquaries of Scotland*, **115**, 159–198.

BENNETT, K. D., BOREHAM, S., SHARP, M. J. & SWITSUR, V. R. 1992. Holocene history of environment, vegetation and human settlement of Catta Ness, Lunnasting, Shetland. *Journal of Ecology*, **80**, 241–73.

BLACKFORD, J. J., EDWARDS, K. J., DUGMORE, A. J., COOK, G. T. & BUCKLAND, P. 1992. Hekla-4, Icelandic volcanic ash and the mid-holocene Scots Pine decline in northern Scotland. *The Holocene*, **2**, 260–265.

BOGAARD, C., DÖRFLER, W., SANDGREN, P. & SCMINCKE, H.-U. 1994. Correlating the Holocene records, Icelandic tephra found in Schleswig-Holstein (Northern Germany). *Naturwissenschaften*, **81**, 1–5.

BRADLEY, R. S. & JONES, P. D. 1992. Records of explosive volcanic eruptions over the last 500 years. *In:* BRADLEY, R. S. & JONES, P. D. (eds) *Climate Since AD 1500*, Routledge, 606–622, London.

BRAYSHAY, M & GRATTAN, J. 1999. Environmental and social responses in Europe to the 1783 eruption of the Laki fissure volcano in Iceland: a consideration of contemporary documentary evidence. *This volume*.

BUCKLAND, P. C., DUGMORE, A. J. & EDWARDS, K. J. 1997. Bronze Age Myths? Volcanic activity and human response in the Mediterranean and North Atlantic Regions. *Antiquity*, **71**, 581–593

——, FOSTER, P., PERRY, D. W. & SAVORY, D. 1981. Tephrochronology and palaeoecolgy, the value of isochrones. *In:* SELF, S. & SPARKS, R. S. J. (eds) *Tephra Studies*. Reidel, 381–389.

BULL, K. 1991. Critical load maps for the U.K. *NERC News* **July**, 31–32.

BURGESS, C. 1985. Population, climate and upland settlement. *British Archaeological Reports*, **143**, 195–229.

—— 1989. Volcanoes, catastrophe and the global crisis of the late second millennium BC. *Current Archaeology*, **117**, 325–329.

CAMUFFO, D. & ENZI, S. 1995. Impacts of clouds of volcanic aerosols in Italy during the last 7 centuries. *Natural Hazards*, **11**, 135–161.

CASELDINE, C., HATTON, J., HUBER, U., CHIVERELL, R. & WOOLEY, N. 1998. Assessing the impact of volcanic activity on mid-holocene climate in Ireland: the need for replicate data. *The Holocene*, **8**, 105–111.

CHARMAN, D. J., GRATTAN, J. P. & WEST, S. 1995. Environmental response to tephra deposition in the Strath of Kildonan, northern Scotland. *Journal of Archaeological Science*, **22**, 799–809.

CHESTER, D. K. 1988. Volcanoes and climate, recent volcanological perspectives. *Progress in Physical Geography*, **12**, 1–35.

DEVINE, J. D., SIGURDSSON, H. DAVIS, A. N. & SELF. S. 1984. Estimates of sulfur and chlorine yield to the atmosphere from volcanic eruptions and potential climatic effects. *Journal of Geophysical Research*, **89**, 6309–6325.

DUGMORE, A. J. 1989. Icelandic volcanic ash in Scotland. *Scottish Geographical Magazine*, **105**, 168–172.

EDWARDS, K. J., BUCKLAND, P. C., BLACKFORD, J. J., DUGMORE, A. J. & SADLER, J. P. 1994. The impact of tephra, proximal and distal studies of Icelandic eruptions. *In:* STÖTTER, J. & WILHELM, F. (eds) *Environmental Change in Iceland*. Münchener Geographische Abhandlungen, **B12**, 79–99.

GOREHAM, E. 1987. Group summary report, wetlands. *In:* HUTCHINSON, T. C. & MEEMA, K. M. (eds) *Effects of Atmospheric Pollutants on Forests, Wetlands and Agricultural Systems*. Springer, 631–636.

GRAF, H.-F., KIRCHNER, I., ROBOCK, A. & SCHULT, I. 1992. *Pinatubo eruption winter climate effects, model versus observations*. Report No. 94, Max Plank Institut für Geomorphologie.

——, ——, —— & —— 1993. Pinatubo eruption winter climate effects, model versus observations. *Climate Dynamics*, **9**, 81–93.

GRATTAN, J. P. 1994. Land abandonment and Icelandic volcanic eruptions in the late 2nd millennium BC. *In:* Stötter, J. & Wilhelm, F. (eds) *Environmental Change in Iceland.* Münchener Geographische Abhandlungen, **B12**, 111–132.

—— & Brayshay, M. B. 1995. An amazing and portentous summer: environmental and social responses in Britain to the 1783 eruption of an Iceland volcano. *The Geographical Journal*, **161**, 125–134.

—— & Charman, D. J. 1994 The palaeoenvironmental implications of documentary evidence for impacts of volcanic volatiles in western Europe, 1783. *The Holocene*, **4**, 101–106.

—— & Gilbertson, D. D. 1994 Acid-loading from Icelandic tephra falling on acidified ecosystems as a key to understanding archaeological and environmental stress in northern and western Britain. *The Journal of Archaeological Science*, **21**, 851–859.

—— & Pyatt, F. B. 1994 Acid damage in Europe caused by the Laki Fissure eruption – an historical review. *The Science of the Total Environment*, **151**, 241–247.

—— & Sadler, J. P. 1999. Regional warming of the lower atmosphere in the wake of volcanic eruptions: the role of the Laki fissure eruption in the hot summer of 1783. *This volume.*

Hall, V. A., Pilcher, J. R. & McCormac, F. G. 1993. Tephra dated lowland landscape history of the north of Ireland, A.D. 750–1150. *New Phytologist*, **125**, 193–202.

——, —— & —— 1994. Icelandic volcanic ash and the mid-Holocene Scots pine (*Pinus sylvestris*) decline in the north of Ireland, no correlation. *The Holocene* **4**, 79–83.

Hammer, C. U. 1977. Past volcanism revealed by Greenland ice sheet impurities. *Nature*, **270**, 482–486.

—— 1984. Traces of Icelandic eruptions in the Greenland ice sheet. *Jökull*, **34**, 51–65.

——, Clausen, H. B. & Dansgaard, W. 1980. Greenland ice sheet evidence of post glacial volcanism and its climatic impact. *Nature*, **288**, 230–235.

——, —— & —— 1981. Past volcanism and climate revealed by Greenland ice cores. *Journal of Volcanology and Geothermal Research*, **11**, 3–10.

——, ——, Friedrich, W. L. & Taubert, H. H. 1987. A Minoan eruption of Santorini in Greece dated to 1645 BC. *Nature*, **328**, 517–519.

Handler, P. 1989. The effect of volcanic aerosols on the global climate. *Journal of Volcanology and Geothermal Research*, **37**, 233–249.

Hansen, J. E. & Lacis, A. A. 1990. Sun and dust versus greenhouse gases, an assessment of their relative roles in global climate change. *Nature*, **346**, 713–719.

Harington, C. R. (ed.) 1992. *The Year Without a Summer.* Canadian Museum of Nature.

Herron, M. M. 1982. Impurity sources of F, Cl, NO and SO in Greenland and Antarctic precipitation. *Journal of Geophysical Research*, **87**, 3052–3060.

Ingram, H. A. P. 1978. Soil layers in mires, function and terminology. *Journal of Soil Science*, 29, 224–227

—— 1987. Ecohydrology of Scottish peatlands. *Transactions of the Royal Society of Edinburgh, Earth Sciences*, **78**, 287–296.

Kelly, P. M., Wigley, T. M. L. & Jones, P. D. 1984. European pressure maps for 1815–16, the time of the eruption of Tambora. *Climate Monitor*, **13**, 76 = –91.

Kerr, A. 1994. Did Pinatubo send climate warming gases into a dither. *Science*, **263**, 1562.

Kington, J. A. 1992. Weather patterns over Europe in 1816. *In:* Harington, C. R. (ed.) *The Year Without a Summer*, Canadian Museum of Nature, 358–377.

Lamb, H. H. 1970. Volcanic dust in the atmosphere; with a chronology and assessment of its meteorological significance. *Philosophical Transactions of the Royal Society, London*, Series **A**, **266**, 425–533.

—— 1977. *Climate, Past, Present and Future.* Volume 2. Methuen.

—— 1988. Volcanoes and climate, an updated assessment. *In:* Lamb, H. H. (ed.) *Weather, Climate and Human Affairs*, Routledge, 301–328.

—— 1992. First essay at reconstructing the general atmospheric circulation in 1816 and the Early Nineteenth Century. *In:* Harington, C. R. (ed.) *The Year Without a Summer*, Canadian Museum of Nature, 355–357.

Lotter, A. F. & Birks, H. J. B. 1993. The impact of the Laacher Sea tephra on terrestrial and aquatic ecosystems in the Black Forest, Southern Germany. *Journal of Quaternary Science*, **8**, 263–276.

LUHR, J. 1992. Volcanic shade causes cooling. *Nature*, **354**, 104.
MANLEY, G. 1974. Central England temperatures, monthly means 1659 to 1973. *Quarterly Journal of the Royal Meteorological Society*, **100**, 389–405.
MASS, C. F. & PORTMAN, D. A. 1989. Major volcanic eruptions and climate, a critical evaluation. *Journal of Climate*, **2**, 566–593.
MCCORMICK, P. M., THOMASON, L. W. & TREPTE, C. R. 1995. Atmospheric effcts of the Mt Pinatubo Eruption. *Nature*, **373**, 399–404
MINNIS, P., HARRISON, E. F., STOWE, L. L., GIBSON, G. G., DENN, F. M., DOELLING, D. R. & SMITH, W. L. 1993. Radiative climate forcing by the Mt. Pinatubo eruption. *Science*, **259**, 1411–1415.
MOSSMAN, R. C. 1896. The meteorology of Edinburgh. *Transactions of the Royal Society of Edinburgh*, **39**, 63–207.
NEUMANN, J. 1990. The 1810s in the Baltic region, 1816 in particular, air temperatures, grain supply and mortality. *Climatic Change*, **17**, 97–120.
—— 1992. The 1810s in the Baltic region, 1816 in particular; air temperatures, grain supply and mortality. *In:* HARINGTON, C.R. (ed.) *The Year Without a Summer*, Canadian Museum of Nature, 392–415.
OSKARSSON, N. 1980. The interaction between volcanic gases and tephra, fluorine adhering to tephra of the 1970 Hekla eruption. *Journal of Volcanology and Geothermal Research*, **8**, 251–266.
PALAIS, J. M. & SIGURDSSON, H. 1989. Petrologic evidence of volatile emissions from major historic and pre-historic volcanic eruptions. *In:* BERGER, A., DICKINSON, R. E. & KIDSON, J. W. (eds). *Understanding Climate Change. Geophysical Monograph 52*. **7**, 31–53.
PILCHER, J. R. 1990, Ecology of subfossil oak woods on peat. *In:* DOYLE, R. G. (ed.) *Ecology and conservation of Irish Peatlands*, Royal Irish Academy, 41–47.
—— & HALL, V. A. 1992. Towards a tephrochronology for the north of Ireland. *The Holocene*, **2**, 255–259
POST, J. D. 1977. *The Last Great Subsistence Crisis of the Western World*. John Hopkins University Press.
RAMPINO, M. R. 1988. Volcanic winters. *Annual Review of Earth and Planetary Science*, **16**, 73–99.
RAYNAL, J. P., VERNET, G. & VIVENT, D. 1999. Des volcans et des hommes depuis le Tardi–Glaciaire en Basse Auvergne (Massif Central, France). *Pact*, in press.
RENFREW, C. 1979. The eruption of Thera and Minoan Crete. *In:* SHEETS, P. D. & GRAYSON, D. K. (eds) *Volcanic Activity and Human Ecology*. Academic Press, 565–585.
ROBOCK, A. & MAO, J. 1992. Winter warming from large volcanic eruptions. *Geophysical Research Letters*, **19**, 2405–2408.
—— 1995. The volcanic signal in surface temperature observations. *Journal of Climate*, **8**, 1086–1103.
ROSE, W. I. 1977. Scavenging of volcanic aerosol by ash, atmospheric and volcanological implications. *Geology*, **5**, 621–624.
SCHNEIDER, S. H. 1994. Detecting climatic change signals. Are there any 'fingerprints'? *Science*, **263**, 341–347.
SCORER, R. S. 1992. Deposition of concentrated pollution at large distance. *Atmospheric Environment*, **26A**, 793–805.
SIGURDSSON, H. 1990. Evidence of volcanic loading of the atmosphere and of climate response. *Palaeogeography, Palaeoclimatology, Palaeoecology*, **89**, 277–289.
SKIBA, U., CRESSER, M. S., DERWENT, R. G. & FUTTY, D. W. 1989. Peat acidification in Scotland. *Nature*, **337**, 68–69.
SMITH, C. M. S., CRESSER, M. S. & MITCHELL, R. D. J. 1993. Sensitivity to acid deposition of dystrophic peat in Great Brtain. *Ambio*, **22**, 22–26.
STOMMEL, H. & STOMMEL, E. 1979. The year without a summer. *Scientific American*, **240**, 134–140.
—— 1983. *Volcano Weather. The Story of 1816, the Year Without a Summer*. Seven Seas.
THORARINSSON, S. 1981. Greetings from Iceland. Ash falls and volcanic aerosols in Scandinavia. *Geografiska Annaler*, **63**, 109–118.
TURCO, R. P., WHITTEN, R. C. & TOON, O. B. 1982. Stratospheric aerosols, observation and

theory. *Reviews of Geophysics and Space Physics*, **20**, 233–279.
——, Toon, O. B., Ackerman, T. P., Pollack, J. B. & Sagan, C. 1983. Nuclear winter, global consequences of multiple nuclear explosions. *Science*, **222**, 1283–1292.
——, ——, ——, & —— 1984. 'Nuclear winter' to be taken seriously. *Nature*, **311**, 307–308.
Wisniewski, J. 1982. The potential acidity associated with dews, frosts and fogs. *Water, Air and Soil Pollution*, **17**, 361–377.
Wood, J. D. 1965. The complicity of climate in the 1816 depression in Dumfriesshire. *Scottish Geographical Magazine*, **81**, 5–17.

Characterization of tephras using magnetic properties: an example from SE Iceland

SILVIA GONZALEZ, JENNIFER M. JONES & DAVID L. WILLIAMS

School of Biological & Earth Sciences, Liverpool John Moores University, Byrom Street, Liverpool L3 3AF, UK.

Abstract: Mineral magnetic techniques have been applied to two different stratigraphic profiles excavated in the Oraefi Volcano area, SE Iceland. The profiles include sequences of soils and tephra layers of inferred age. Each profile includes the distinctive pumice tephra layer associated with the eruption of Oraefi in 1362, which is a stratigraphic marker for the area. Detailed magnetic measurements included: magnetic susceptibility, anhysteretic remanent magnetization, isothermal remanent magnetization (IRM) acquisition, back-field IRM's and strong field thermomagnetic behaviour (Curie temperature). The results indicate that the most useful magnetic properties for distinguishing between soils and tephras are the back-field IRM's and the Curie temperatures. Principal component analysis has been applied using all the measured magnetic properties. In this way it is possible to discriminate between different types of material in the sections, and even more importantly to correlate some tephras between the two sections. The potential of this technique to correlate tephras is promising, but further evaluation is required.

The term 'tephra' as defined by Thorarinsson (1944, 1958) applies to 'all the clastic volcanic material which during an eruption is transported from the crater through the air...'. It is a collective term for all airborne pyroclasts including both air-fall and flow pyroclastic material. At present the term tephra has become essentially the equivalent of 'pyroclastic material' without reference to fragment sizes, and it also includes all subaqueous varieties, which eliminates the necessity of interpreting the environment of deposition before using the term (Fisher & Schmincke 1984). The presence and identification of tephra layers in outcrops or substrates led to the development of tephrochronology, the term coined by Thorarinsson (1944), i.e. the derivation of a chronology on the basis of measurement and characterization of successive tephra layers. The presence of clearly identifiable isochrons is of considerable value to the reconstruction of palaeoenvironments. However, as with any chronological marker, the technique relies on the ability to apportion the tephra to a particular event with confidence. It is essential, therefore, to identify those diagnostic properties of tephra deposits which lend themselves readily to such rigorous requirements.

Iceland proved an ideal location for the initial development of tephrochronological studies since it is highly volcanic and contains good archaeo-documentary records for the timing of eruptions from major volcanoes including Hekla, Katla and Oraefi. To date, most published examples of tephrochronological (and tephrostratigraphical) studies report tephra characterization based on colour and particle size characteristics (Thorarinsson 1944) or petrographic and petrochemical analyses (Westgate & Fulton 1975; Smith & Leeman 1982; Hodder *et al.* 1991; Dugmore & Newton 1992; Dugmore *et al.* 1992).

The presence of a magnetic component in many tephra deposits has been recognized for some time. Several authors report the presence of glass-encased magnetite grains in tephra layers (Brewster & Barnett 1979). In addition, a ferrimagnetic component of tephra deposits has been implicated in magnetostratigraphical studies of lacustrine and marine sediments. Such studies report peaks in magnetic susceptibility values coincident with tephra layers in lake sediments in New Guinea (Oldfield *et al.* 1980, 1983), Mexico (Lozano-Garcia & Ortega-Guerrero 1994) and North Atlantic marine sediments (Robinson 1986) amongst others. Recent studies of the magnetic characteristics of Icelandic tephra include the magnetostratigraphical study of Lake Vatnsdalsvatn sediments (cited in Thompson & Oldfield 1986) and the analysis of the sediments of Lake Svinavatn, north of Iceland (Bradshaw & Thompson 1985).

The magnetic characteristics of tephra will reflect the chemical composition of the source material together with the temperature regime of the eruptive event and subsequent chilling of the magma. If magnetic characterization of tephra is sufficiently discriminating it would provide a valuable tool for soil chronosequence studies by permitting accurate identification of individual tephra marker horizons. In addition, it would allow identification of tephras where it is not possible to identify them on field evidence alone. For example, in those situations where there is difficulty discriminating between, say, soil and tephra. Some attempts have been made to evaluate the discriminating potential of magnetic measurements for tephra identification and correlation (Urrutia-Fucugauchi 1990; Robertson 1993); however few studies have combined empirical studies with multivariate statistical analysis to identify the magnetic signature of individual tephras. The aim of this study is to determine the mineral magnetic characteristics of soils and tephras from two sites in SE Iceland and to evaluate the discriminant capacity of such data for the identification of tephras from spatially and temporally separated sources.

Site information

The study area is located in southeast Iceland, near the western outlet of the Oraefi glaciers (Fig. 1). The field work was carried out on the 'Joint Liverpool Iceland Expeditions' of Liverpool Polytechnic (now Liverpool John Moores University) and the University of Liverpool, during the 1987 and 1989 field seasons. Two different stratigraphic profiles were excavated, one in the Storalda moraine and the other in the Skaftafellsheidi (Fig. 2).

The Storalda moraine complex forms the outermost part of the recessional moraines of the Svinafellsjokull. The Storalda profile was located on the penultimate distal ridge of the complex and was excavated to a depth of 1.50 m. (Fig. 3). The uppermost 40 cm of the profile comprised soil with 3 clearly defined tephra horizons. From 40–130 cm a loess-type material was found with 5 tephra horizons, the lowest of which is a white rhyolitic pumice with large fragments, and contrasts with the other tephras which are darker and finer, and have basic compositions. The darker tephras are associated with the Katla and Laki volcanoes (Fig. 1), whilst the white layer is characteristic of the 1362 Oraefi tephra, which has been used as a stratigraphic marker for the SE of Iceland since the work of

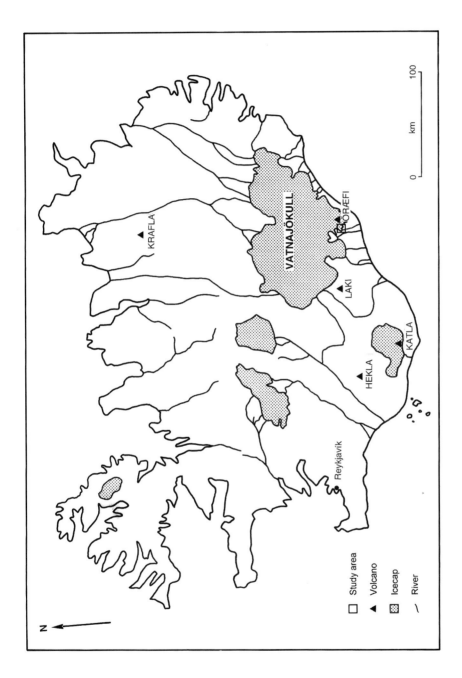

Fig. 1. Location of the study area in the SE of Iceland, the black triangles mark the positions of active volcanos which have produced tephras found in the study area.

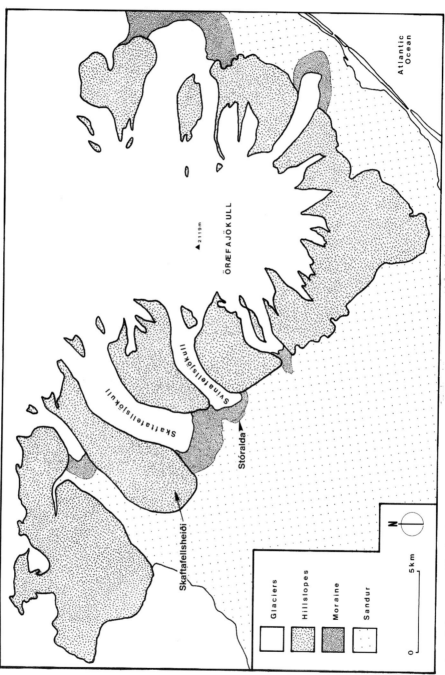

Fig. 2. Location of the two profiles in the Oraefi area: Storalda moraine and Skaftafellsheidi. The black triangle shows the position of the Oraefajokull volcano.

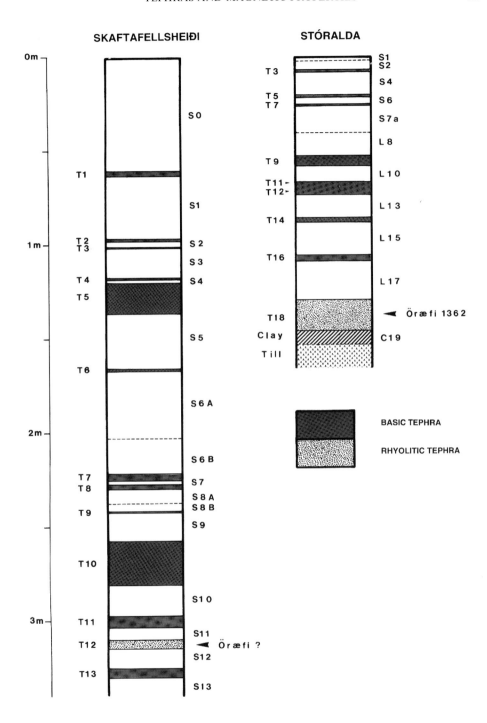

Fig. 3. Profiles excavated in the Storalda moraine and Skaftafellsheidi, showing the tephra layers (T) and associated soils.(S = soil, L = loess). Ages of the tephras according to historical records (Thorarinsson 1958).

Thorarinsson (1958). The entire profile yielded 5 soil samples, 5 loess and 8 tephra samples. The upper part of the fifth tephra was 'disturbed' in such a way that it was uncertain whether it represented another tephra so a sample was taken at this level (T11) and another in the main body of the layer (T12).

At the Skaftafellsheidi site, samples were obtained from an exposed section in a stream gully. This section features 13 different soil horizons with 13 intercalated tephra horizons (Fig. 3). The soils at 170–200 cm showed some subdivision on the basis of their colour and, therefore, were subdivided into samples with suffixes 'A' and 'B'. The tephras T11 and T12 have a rhyolitic composition and light colours, and T12 is correlated with the 1362 Oraefi tephra. The rest of the tephras in the profile are of more basic compositions (black colour) associated with the eruptions of the Katla, Laki and Krafla volcanoes. Profile diagrams for both the Storalda and Skaftafellsheidi sites are shown in Fig. 3, indicating the composition of the tephras.

Magnetic properties

The type and concentration of magnetic minerals in a tephra depends in the first instance on the chemical composition; for example, in basaltic compositions with high contents of FeO and TiO_2 the quantity of magnetic minerals is higher than in rhyolitic compositions. The determination of the magnetic properties in a tephra gives very useful information about the type, concentration and grain size of the magnetic minerals. In this way it is possible to make inferences about the chemical composition, mechanism of emplacement, cooling rates, etc. in the tephras. On the other hand, the soils produced from the weathering of the tephras, will have different magnetic mineralogy, because the primary magnetic minerals are altered and new minerals are formed during the process of pedogenesis.

The main advantages of using magnetic properties to differentiate and correlate tephra layers are that the measurements are quick, easy and economical (especially in comparison with certain geochemical methods), allowing the processing of a large number of samples in a short time.

Sample preparation

From each distinct layer of tephra or soil identified in the two profiles studied, a representative bulk sample was taken in the field. In total 46 samples were collected for analysis. In the laboratory the samples were air dried for several days, gently disaggregated with a porcelain pestle and mortar and a known weight of each sample was tightly packed in a 10 cm^3 polystyrene non-magnetic container. Although several workers have recommended the use of a particular grain size for magnetic measurements (Oldfield et al. 1983; Oldfield 1991) the <2 mm fraction was measured in this case to allow direct comparisons with bulk geochemical analyses.

Magnetic measurements

The prepared samples were subjected to a series of magnetic measurements in the following order (See Table 1 for an explanation of the meaning of each magnetic property):

Table 1. *Glossary of magnetic properties measured and their interpretation (modified from Thompson & Oldfield 1986; Walden et al. 1992).*

Magnetic susceptibility (X)	A measure of the degree to which a substance can be magnetised. Its value is roughly proportional to the concentration of ferrimagnetic and paramagnetic minerals within the sample.
Frequency dependent susceptibility (Xfd%)	Measures the variation of magnetic susceptibility with the frequency of the applied alternating magnetic field. Its value is proportional to the amount of 'viscous' magnetic grains which lie at the stable single domain/superparamagnetic boundary, where these grains show a delayed response to the magnetizing field. The parameter is expressed as a percentage difference between low (Xlf) and high frequency (Xhf) susceptibility.
Anhysteretic remanent magnetization (ARM)	The remanence produced in a sample by subjecting it to an increasing and decreasing (0 mT–100 mT–0 mT) alternating magnetic field superimposed upon a steady dc magnetic field (0.04 mT). The remanence is proportional to the steady field. Can be measured on a constant volume or mass-specific basis.
Isothermal remanent magnetization (IRM)	The remanence produced in a sample by the application and subsequent removal of a known magnetic field; the magnetization grows parallel to the applied field.
Saturation isothermal remanent magnetization (SIRM)	The highest amount of magnetic remanence that can be produced in and retained by a sample as a result of its emplacement in a strong magnetic field at a given temperature (usually room temperature). Can be measured on a constant volume or mass-specific basis.
Back-field ratios	A sample containing SIRM (given along the +x axis) is then given a series of IRM's in the opposite direction (−x). The change in magnetization is expressed as a percentage of the difference in magnetization between the original SIRM in the +x direction and a SIRM in the −x direction. Given in a scale of +1 to −1 or 0–100%. The ratios give some indication of ferrimagnetic (magnetite) versus canted antiferromagnetic (hematite, goethite) minerals within a selection of samples.
Curie temperature (Tc)	The temperature above which a ferromagnetic or ferrimagnetic substance loses its magnetization, becoming paramagnetic (Curie point). This type of measurement aids the identification of the magnetic minerals present in a sample, because the mineral composition controls the Curie temperature.

(a) **Magnetic susceptibility (X)**, using a Bartington MS2 susceptibility meter with a dual frequency sensor, measuring high (Xhf) and low frequencies (Xlf), to calculate **frequency dependent susceptibility** (Xfd%).

(b) **Anhysteretic remanent magnetization (ARM)** was given using a Molspin alternating demagnetizer with a specially designed direct field coil. The sample was subjected to an alternating field of 100 mT, superimposed upon a constant dc field of 0.04 mT (similar to the earth's magnetic field in the UK at the present). The resulting remanent magnetizations were measured on a Minispin magnetometer.

(c) **Isothermal remanent magnetizations (IRM)** were given by exposing the samples to a range of fields, produced with 2 different Molspin pulse magnetizers (0–300 mT and 0–1000 mT). The resulting remanences were measured using a Minispin magnetometer.

(d) **Saturation isothermal remanent magnetization (SIRM)** was calculated, where SIRM = IRM at 1000 mT. This field is sufficient to give a SIRM to most magnetic minerals, however, haematite and goethite will not be saturated at

1000 mT, therefore the term 'saturation IRM' is not strictly accurate in all cases; in addition, the 'Soft' IRM (=SIRM −IRM in a field of −20 mT) and the 'Hard' IRM (=SIRM+IRM in a field of −300 mT) where calculated.

(e) **Back-field ratios:** The sample containing a SIRM in +x direction was exposed to a series of reverse magnetic fields along the -x direction with a Molspin pulse magnetizer (in this study the fields were −20 mT, −40 mT, −100 mT and −300 mT), measuring the resulting magnetic remanence in a Molspin magnetometer at each step.

(f) **Curie temperatures:** the strong field thermomagnetic behaviour was determined using a horizontal Curie balance. The measurements were made in air, with a magnetic field of $c.$ 0.5 T, heating the sample up to 700°C in 15 minutes and then cooling to room temperature.

Results and interpretation

The results of the magnetic properties and parameters calculated for the samples are presented in Figs 4 to 9, where they are plotted against the depth in the profiles.

The magnetic property which has been widely used to try to establish correlations between tephras is magnetic susceptibility (Thompson & Oldfield 1986; Urrutia-Fucugauchi 1990), where correlations have been made trying to match the 'peaks' in the susceptibility associated with the presence of tephras. However, from Fig. 4 it is obvious that a simple visual matching is not possible in this case. The reason for this could either be because the same number of tephras are not represented in the 2 profiles studied or/and strong changes in the grain size for the same tephra layer occur within a short distance.

The same occurs for the determinations of ARM, IRM, SIRM, where again a simple visual correlation is not possible (Figs 5, 6, 7). For this reason it was decided to use a numerical approach, applying a series of multivariate statistical tests to see if in a more objective way, it is possible to correlate tephras (see the next section for the description of the results).

On the other hand, and despite all the problems previously mentioned, the graphs of the back-field IRMs (Fig. 8) show interesting results differentiating the presence of soils and tephras in the profiles. Whilst in the case of the black, basic tephras the graphs show peaks towards the left (e.g. T3, T16, Storalda profile), the loessic soils, organic soils and the white rhyolitic tephras have peaks to the right (e.g. S7, T12, Skaftafellsheidi section). This indicates that the soils and rhyolitic tephras contain more low coercivity magnetic minerals such as multi-domain magnetite, compared to the basic tephras which seem to contain more minerals with intermediate coercivities (100–300 mT). In this way the back-field IRM ratios can be used as an easy and quick alternative way to discriminate between tephras and soils, and even between different compositions for the tephras.

The determinations of the Curie temperatures show strong differences in the form of the thermomagnetic curves between soils and tephras (Figs 9a & b). The curves in the soils are generally more reversible, showing fewer changes in the mineralogy between the cycles of heating and cooling. The tephras, on the other hand, in general show, irreversible curves, with strong changes in the mineralogy during the cycles of heating and cooling.

From the shape of the Curie curves, which show no obvious Curie temperature,

TEPHRAS AND MAGNETIC PROPERTIES

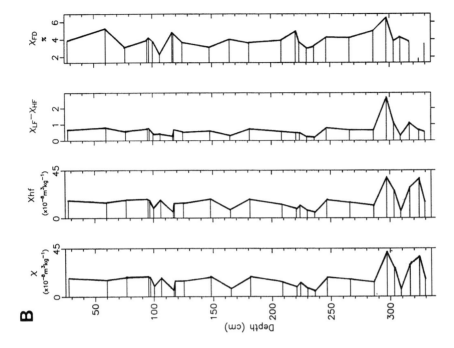

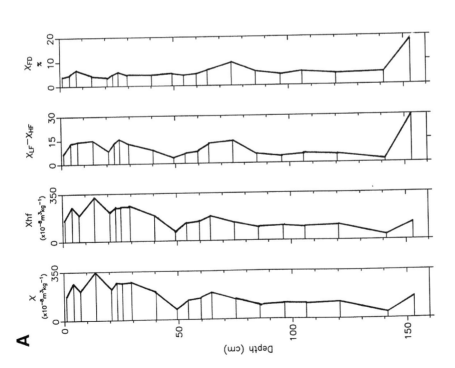

Fig. 4. The results of the measurement of magnetic susceptibility (X), X_{HF} (susceptibility at high frequency), X_{LF} (susceptibility at low frequency), X_{FD} (frequency dependence), in the profiles at: (**A**) Storalda and (**B**) Skaftafellsheidi.

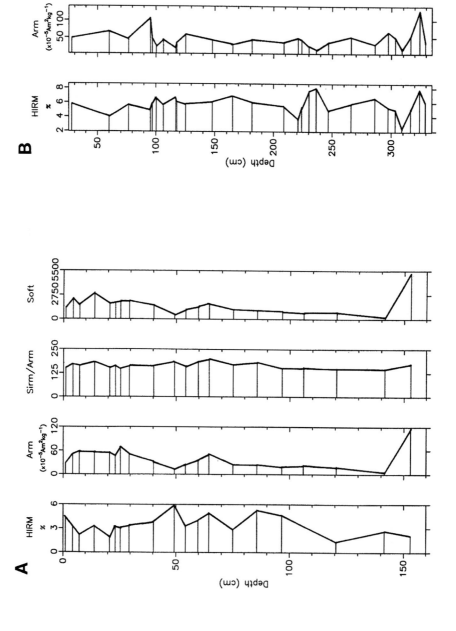

Fig. 5. Results obtained in (**A**) Storalda and (**B**) Skaftafellsheidi for: the HIRM % (hard IRM), ARM (anhysteretic remanent magnetization), the ratio between SIRM/ARM and the soft IRM. For an explanation of the meaning of these parameters see Table 1 and the text.

TEPHRAS AND MAGNETIC PROPERTIES 135

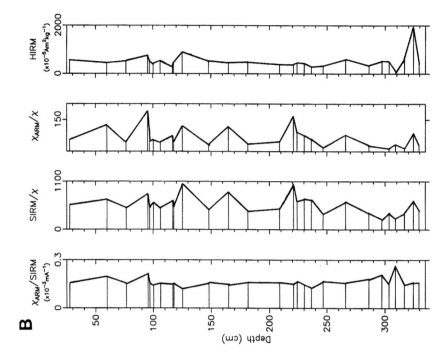

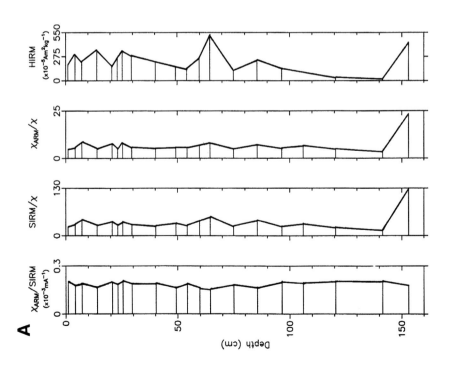

Fig. 6. The variation with depth in the two profiles, (**A**) Storalda and (**B**) Skaftafellsheidi, of the following ratios: $X_{ARM}/SIRM$, $SIRM/X$, X_{ARM}/X, HIRM (hard IRM).

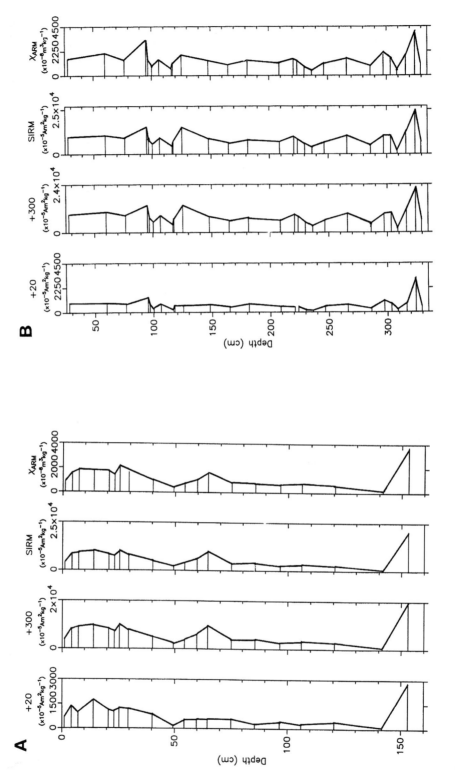

Fig. 7. The variation with depth in the two profiles, (A) Storalda and (B) Skaftafellsheidi, of the following parameters: IRM at +20 mT, IRM at +300 mT, SIRM (saturation isothermal remanent magnetization) and X_{ARM}.

TEPHRAS AND MAGNETIC PROPERTIES 137

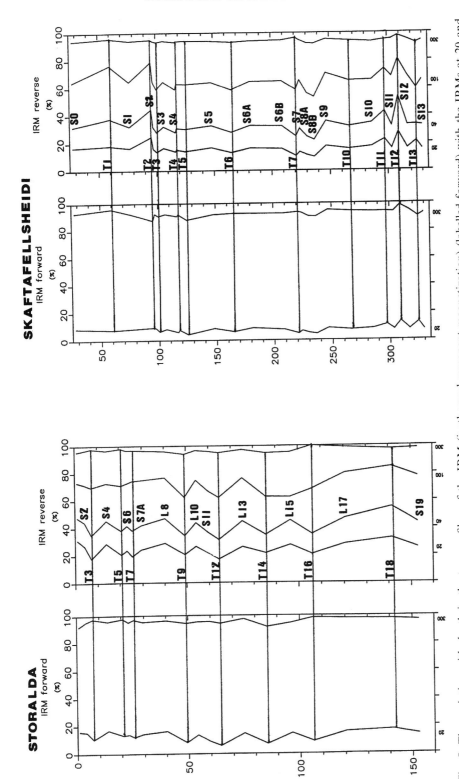

Fig. 8. The variation with depth in the two profiles, of the IRM (isothermal remanent magnetization) (labelled forward) with the IRMs at 20 and 300 mT expressed as percentages of the SIRM and reverse (back-field) IRM at fields of 20, 40, 100 & 300 mT (see Table 1 for further details). T = tephras, S = soils and L = loessic soils. The sample numbers are the same as in Fig. 3. See discussion of the results in the text.

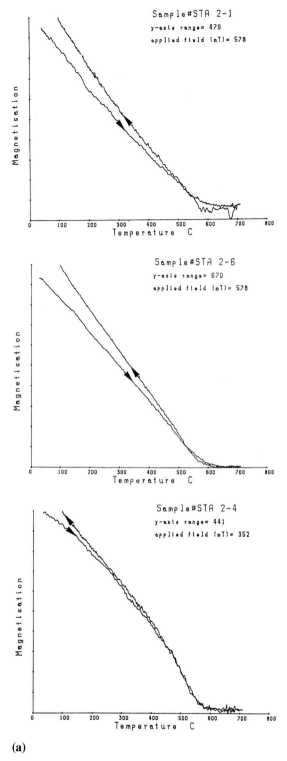

(a)

Fig. 9. Some examples of thermomagnetic curves of (a) soils of the Storalda profile, showing reversible curves associated with the presence of magnetite, with a Curie temperature of c. 580°C.

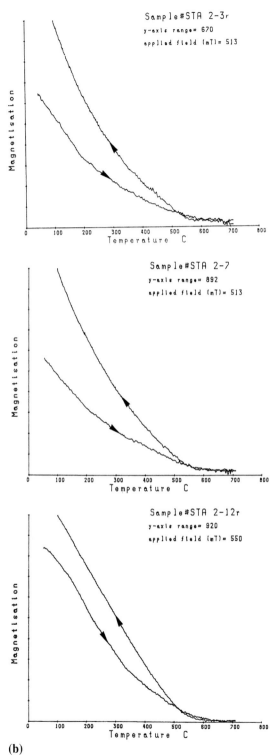

Fig. 9. (**b**) Some examples of thermomagnetic curves of tephras of the Storalda profile, showing irreversible curves without a clear Curie point, associated with the presence of paramagnetic minerals.

the tephras seem to contain predominantly paramagnetic minerals. On the other hand for the soils the curves are more typical of ferrimagnetic minerals, and Curie temperatures of around 570–580°C indicate the presence of magnetite.

Apparently of all the magnetic properties measured during this study, the magnetic properties which discriminate best between soils and tephras are the backfield IRM ratios and the Curie temperatures, especially if a combination of the two properties is used.

Statistical analysis

From the above data, it is clear that it is possible to differentiate relatively easily between soils and tephras using their magnetic properties, but not to correlate between a single tephra layer, using only visual matching of the graphs obtained for each profile. For this reason multivariate statistical analysis was employed, in order to assess in a more objective way the potential for distinguishing and correlating tephra and soil horizons using their magnetic properties.

Principal component analysis (PCA) numerically describes relationships between a large number of correlated variables in terms of a smaller set of new, independent variables or principal components.

PCA (see Manly 1986; Ferguson 1994 for a description of the method) was applied using the PATN program of Belbin (1987). The program considers all the magnetic properties measured and parameters previously calculated and compares the multivariate statistics of the 2 different datasets obtained from the Storalda and Skaftafellsheidi profiles, to identify ultimately groups of samples with similar magnetic characteristics. A total of 13 different magnetic properties/parameters were taken into consideration: Xlf, Xfd, SIRM, HIRM, ARM, Soft IRM, FD%, Xarm/S, Xarm/X, $-20\,mT\%$, $-40\,mT\%$, $-100\,mT\%$, $-300\,mT\%$. The raw data matrix used during the analysis consisted of 47 samples (from the 2 sites) by 13 variables.

Principal component analysis (PCA), for example, has been applied often and successfully to pollen stratigraphic data for the precise recognition, description and characterization of pollen assemblage zones (Birks & Berglund 1979); it is believed to be equally applicable to the zonation of other types of palaeoecological sequences, such as diatoms or sediment chemistry (Pennington & Sackin 1975), and in studies of provenance of sediments using environmental magnetism (Walden et al. 1991, 1992; Thompson et al. 1992).

The results for the Storalda profile (Fig. 10) demonstrate that the PCA diagram discriminates well, showing different areas for the black (basic) tephras of the Katla and Laki eruptions, the Oraefi 1362 white (rhyolitic) tephra, loess, organic soils and till.

The results for the Skaftafellsheidi profile (Fig. 11) show the same type of correlation, discriminating again between different tephra compositions and soils, apparently with similar positions in the diagram for the black tephras and soils as in the Storalda profile.

However, in Fig 11 it is not sufficiently clear whether T11 or T12 correspond to the Oraefi 1362 tephra or not. For this reason, it was necessary to put the 2 profiles in the same diagram (Fig. 12) to try to compare the 2 profiles under the same conditions. The results indicate that the tephra STT18 of the Storalda profile has a very similar magnetic signal to the tephra SKT12 of the Skaftafellsheidi profile; for

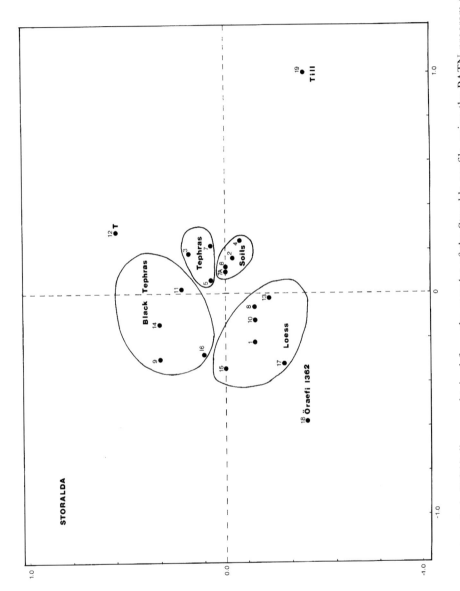

Fig. 10. Principal component analysis (PCA) diagram obtained from the samples of the Storalda profile, using the PATN program (Belbin 1987), considering all the magnetic properties/parameters measured during this study. See discussion in the text.

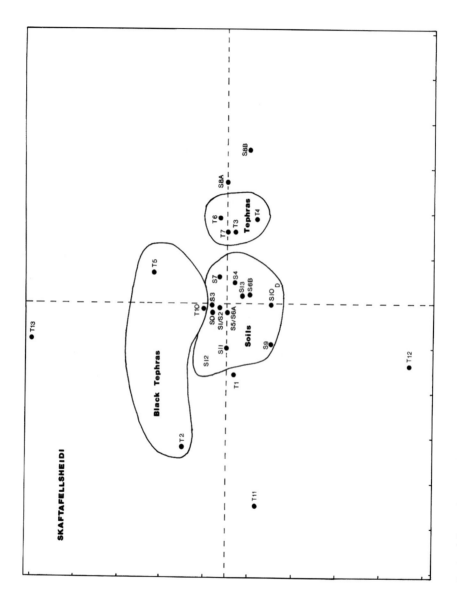

Fig. 11. PCA diagram obtained from the samples of the Skaftafellsheidi profile, using the PATN program (Belbin 1987), considering all the magnetic properties measured during this study.

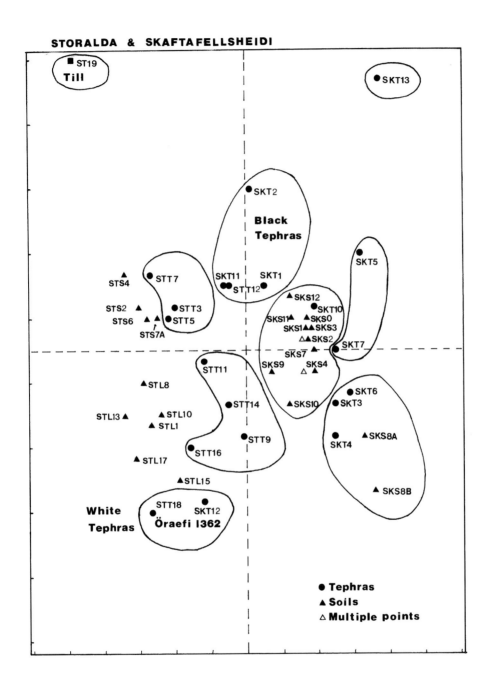

Fig. 12. PCA diagram considering all the magnetic properties measured during this study, putting together the two studied profiles, differentiating between tephras of different compositions (basic, intermediate, acid), soils, till.
ST = Storalda sample, SK = Skaftafellsheidi sample.

this reason it is believed that they are the same tephra.

From Fig. 12, it would seem that a clear multivariate pattern exists in the mineral magnetic data for the samples analysed, making stratigraphic correlations possible between particular tephra horizons and derived soils. It is interesting to note that the PCA program shows only one group containing both tephras and soils (samples SKT3, SKT4, SKT6, SKS8A and SKS8B). This may indicate possible contamination due to mixing during sampling, especially when dealing with very thin tephra horizons. The results reported here are encouraging so far, showing that it does seem possible to characterize tephras and soils using their magnetic properties.

Conclusions

Detailed measurements of the magnetic properties in 2 different stratigraphic profiles in the SE of Iceland in the Oraefi volcano area, show that these techniques are potentially very useful for discriminating and correlating tephras and soils.

The magnetic properties which discriminate better between tephras and soils are:

(a) the back-field isothermal remanent magnetizations (IRMs), showing the presence of different types of magnetic minerals for the soils (magnetite) and the tephras (paramagnetic minerals);

(b) the determination of the Curie temperatures, using the form of the curves to distinguish between the tephras (irreversible curves with a high contribution of paramagnetic minerals) and the soils (more reversible curves, with Curie temperatures between 570–590°C, indicating the presence of magnetite).

Correlation between some tephras in the two profiles was achieved applying principal component analysis (PCA), using all the measured magnetic properties and parameters calculated during this study. Using PCA it is possible to differentiate apparently between different compositions of tephras (rhyolitic, intermediate and basic) and different types of soils (loessic, organic).

Despite the limited dataset the results are encouraging so far, showing that it seems to be possible to characterize different types of tephra and soil horizons using magnetic 'fingerprinting'. The combination of the use of detailed magnetic measurements together with the analysis of the information using PCA to correlate tephras and soils is very promising, but more work and further evaluation is necessary to assess their complete potential.

The authors would like to thank Dr Alan Thompson, Jan Hookey and Patrick Bonnet for their technical assistance in the field. Also thanked are Ragnar Stefánsson and family, especially Anna-Maria for their help. The Icelandic Research Council are thanked for permission to work in Iceland and Stefán Benediktsson the director of the Skaftafell National Park. The authors would also like to thank Prof. Frank Oldfield and Bob Jude from the Geography Department of Liverpool University for access to some of the mineral magnetic instrumentation and PATN program. Also thanked are Prof. John Shaw and Dr Tim Rolph from the Geomagnetism Laboratory of Liverpool University for the use of the Curie balance and helpful discussions during this work.

References

BELBIN, L. 1987. *Pattern Analysis Package (PATN)*. CSIRO, Division of Water and Land Resources, PO Box 1666, Canberra, ACT, Australia 2601.

BIRKS, H. J. B. & BERGLUND, B. E. (1979). Holocene pollen stratigraphy of southern Sweden: a reappraisal using numerical methods. *Boreas*, **8**, 257–279.

BRADSHAW, R. & THOMPSON, R. 1985. The use of magnetic measurements to investigate the mineralogy of some Icelandic lake sediments and to study catchment processes. *Boreas*, **14**, 203–215.

BREWSTER, G. R. & BARNETT, R. L. 1979. Magnetites from a new unidentified tephra source, Banff National Park, Alberta. *Canadian Journal of Earth Sciences*, **16**, 1294–1297.

DUGMORE, A. J. & NEWTON, A. J. 1992. Thin tephra layers in peat revealed by X-radiography. *Journal Archaeological Science*, **19**, 163–170.

——, —— & SUGDEN, D. 1992. Geochemical stability of fine-grained silicic Holocene tephra in Iceland and Scotland. *Journal of Quaternary Science*, **7**(2), 173–183.

FERGUSON, J. 1994. *Introduction to Linear Algebra in Geology*. Chapman & Hall, London.

FISHER, R. V. & Schmincke H. U. 1984. *Pyroclastic Rocks*, Springer-Verlag, Berlin.

HODDER, A. P. W., DE LANGE, P. J. & LOWE, D. J. 1991. Dissolution and depletion of ferromagnesian minerals from Holocene tephra layers in an acid bog, New Zealand, and implications for tephra correlation. *Journal of Quaternary Science*, **6**(3), 195–208.

LOZANO-GARCIA, M. S. & ORTEGA-GUERRERO, B. 1994. Palynological and magnetic susceptibility records of Lake Chalco, Central Mexico. *Palaeogeography, Palaeoclimatology, Palaeoecology*, **109**, 177–191.

MANLY, B. F. J. 1986. *Multivariate Statistical Methods: A Primer*. Chapman & Hall, London.

OLDFIELD, F. 1991. Environmental Magnetism; a personal perspective. *Quaternary Science Reviews*, **10**(1), 73–85.

——, APPLEBY, P. G. & THOMPSON, R. 1980. Palaeoecological studies of three lakes in the Highlands of Papua New Guinea. I. The chronology of sedimentation. *Journal of Ecology*, **68**, 457–477.

——, BATTARBEE, R. & DEARING, J. A. 1983. New approaches to recent environmental change. *Geographical Journal*, **149**, 167–181.

PENNINGTON, W. & SACKIN, M. J. 1975. An application of principal component analysis to the zonation of two Late-Devensian profiles I. Numerical analysis. *New Phytologist*, **75**, 419–440.

ROBERTSON, D. 1993. Discrimination of tephra using rockmagnetic characteristics. *Journal of Geomagnetism and Geoelectricity*, **45**, 167–178.

ROBINSON, S. G. 1986. The late Pleistocene palaeoclimatic record of North Atlantic deep-sea sediments revealed by mineral-magnetic measurements. *Physics of the Earth and Planetary Interiors*, **42**, 22–57.

SMITH, D. R. & LEEMAN, W. P 1982. Mineralogy and phase chemistry of Mount St. Helens tephra sets W & Y as keys to their identification. *Quaternary Research*, **17**, 211–227.

THOMPSON, R. & OLDFIELD, F. (1986) *Environmental Magnetism*. Allen & Unwin, London.

——, CAMERON, C., SCHWARZ, K. A., JENSEN, V., MAENHAUT VAN LEMBERGE & L. P. SHA 1992. The magnetic properties of Quaternary and Tertiary sediments in the southern North Sea. *Journal of Quaternary Science*, **7**, 319–334.

THORARINSSON, S. 1944. Tefrokronologiska studier pa Island. *Geografiska Annaler*, **26**, 1–217.

—— 1958. The Oraefajokull eruption of 1362. *Acta Naturalia Islandica*, **2**.

URRUTIA-FUCUGAUCHI, J. 1990. Characterization and correlation of tephra from their magnetic properties. *(EOS), Transactions of the American Geophysical Union*, **71**(43), 1283 (abstr.).

WALDEN, J., SMITH, J. P. & DACKOMBE, R. V. (1992) Mineral magnetic analyses as a means of lithostratigraphic correlation and provenance indication of glacial diamicts: intra-and inter-unit variation. *Journal of Quaternary Science*, **7**(3), 257–270.

——, WHITTAKER, R. J. & HILL J. 1991. The use of mineral magnetic analyses as an aid in investigating the recent volcanic disturbance history of the Krakatau Islands, Indonesia. *The Holocene*, **1**(3), 262–268.

WESTGATE, J. A. & FULTON, R. J. 1975. Tephrostratigraphy of Olympia Interglacial sediments in south-central British Columbia, Canada. *Canadian Journal of Earth Science*, **12**, 489–502.

An assessment of discriminant function analysis in the identification and correlation of distal Icelandic tephras in the British Isles

DANIEL J. CHARMAN & JOHN GRATTAN[1]

The University of Plymouth, Department of Geographical Sciences, Drake Circus, Plymouth PL4 8AA, UK
[1] *Present address: University of Wales, Institute of Geology and Earth Studies, Llandinan Building, Aberystwyth SY23 3DB, UK*

Abstract: This paper assesses the potential of discriminant function analysis (DFA) for tephrochronology in the UK and Ireland. Current identification and correlation of Holocene tephras relies largely on radiocarbon dating to suggest a likely candidate eruption followed by a geochemical comparison using binary and ternary plots of selected major oxides. As more tephras are discovered, and the patterns of deposition appear increasingly complex, this approach is likely to work less effectively. In addition, the utility of tephra for the establishment of chronozones will be limited by the availability of radiometric dates to constrain the initial candidate eruption. The results of a DFA on some of the limited published geochemical data are presented and it is clear that this statistical technique offers advantages to the application and development of tephrochronology in western Europe. Future work should concentrate on the provision of discriminant functions from a more complete reference dataset, which will enable the identification of unknown tephras with a known probability of misclassification.

Tephrochronology has offered a unique relative dating technique in many parts of the world for some years (Keller 1981; Rabek *et al.* 1985; Lowe 1988; Sullivan 1988; Braitseva & Meleketsev 1990; St. Seymour & Kristianis 1995), but one of the main limitations to its utility in the British Isles is the fact that, until recently, clearly defined Icelandic tephra layers were only found over relatively restricted regions, often in relative proximity to the volcano (Larsen & Thórarinsson 1977; Dugmore & Buckland 1991). Though ash-falls had been described and located in Scandinavia and The Faeroes (Persson 1971; Thórarinsson 1980) it was only recently that it became possible to use tephrochronology in the British Isles. Discoveries of tephra from Icelandic sources in British and Irish sediments have demonstrated a clear potential for precise relative dating of Holocene lake sediments and peats (Dugmore 1989; Bennett *et al.* 1992; Blackford *et al.* 1992; Dugmore & Newton 1992; Pilcher & Hall 1992; Hall *et al.* 1993, 1994; Charman *et al.* 1995). Furthermore, recent dating of some of these volcanic events by high precision radiocarbon measurements (Pilcher *et al.* 1995) has shown that it is possible to establish precise calendrical dates for these tephras which will then allow correlation with other proxy records as diverse as tree-ring chronologies (Pilcher *et al.* 1984), ice-core records (Beget *et al.*

1992) and deep sea sediments (Long & Morton 1987; Graham *et al.* 1990) and establish a precisely dated isochrone across wide regions (Buckland *et al.* 1981).

The potential for the use of tephrochronology for Holocene sediments in Britain and Ireland is clearly immense, but depends on adequate identification and correlation of the tephras discovered. Four factors are likely to make this an increasingly difficult task:

1. tephra has now been found from at least eight separate events (Dugmore, pers. comm.) and as more work is carried out by an increasing number of researchers, this number is likely to increase still further;
2. a number of tephras occur quite close together in time, e.g. Hekla 1 AD1104, Oraefajökull AD1362 and a further unknown tephra from AD860 were all found in Sluggan Bog, Northern Ireland (Pilcher *et al.* 1995).
3. the distribution of tephra from any one eruption will not have been even (Thórarinnson 1967) and the same sequence of tephras may not be present at all locations;
4. correlations are likely to be further clouded by the fact that many of the tephras are probably derived from the Hekla volcanic system and thus geochemical signatures, based on a determination of major oxides, are often similar to one another. Thus, the growing body of information has introduced a need for theoretical rigour in tephrochronological studies not considered necessary in studies of rare events (Stanley & Sheng 1986; Sullivan 1988; Rose & Chesner 1990).

At present, the technique for the identification and correlation of a tephra in this region is based on an initial age estimate using radiocarbon, followed by comparison of geochemistry with reference data from a suspected source. The comparison is normally made using mean and standard deviations of major oxides, plus a graphical comparison of the data using bivariate and ternary plots. So far, in most cases, this has worked adequately but, given the factors above, it may not continue to perform so well in the future. There are also other disadvantages to this approach. Firstly, the full complement of geochemical information is not properly used. Only a few selected oxides can be compared at once and even if multiple plots are used it is not easy to assess differences in the geochemical composition. Secondly, it relies on radiocarbon age estimates to provide an initial likely candidate eruption. Besides the inherent circularity in this method, any such guess is likely to be increasingly inaccurate with temporally closely spaced eruptions. It also means that the full utility of tephrochronology as a dating tool is not realized, as profiles will still require a high input of radiocarbon analyses to be sure of identifying a candidate eruption. Thirdly, it is a subjective comparison and cannot provide a quantitative assessment of the best discriminating oxides or of the probability of correct identification of a given tephra. Finally, given the discussion over analytical techniques in micro-analysis of tephra shards (Hunt & Hill 1993, 1994; Bennett 1994), and the difficulty of comparing data generated by different operators on different machines, the robustness of such subjective comparisons may be limited. Clearly while every effort should be made to follow established guidelines in analytical techniques (Froggatt 1992), it is inevitable that differences will still exist between datasets which need to be compared with one another. In the light of these problems, and following Bennett's (1994) suggestion that more sophisticated data analysis could be usefully pursued, this paper aims to assess the potential of one technique – discriminant function analysis (DFA) – for correlating distal Icelandic tephras in the British Isles. This is achieved by the analysis of a small pilot dataset of published data. Further work on a comprehensive dataset will be necessary to provide a definitive set of criteria for the separation of major tephras in the region.

Data assembly and analysis

The application of DFA to geochemical data to discriminate between tephras is not new. Borchardt et al. (1972) first introduced the technique, and it has recently been applied to data from New Zealand (Stokes & Lowe 1988; Stokes et al. 1992) and North America (Shane & Froggatt 1994) with success. The technique is described fully by Stokes & Lowe (1988) and the description here will be brief. The analysis consists of two main parts. Firstly, a classification model is generated from a reference set of data. This dataset contains the major oxides from a series of known tephras, in this case from some of the main Icelandic eruptions thought to be represented in the British Isles. This classification model consists of a series of 'discriminant functions', which are linear combinations of the major oxides. The difference between each known tephra can be measured by the Mahalanobis distance squared statistic and the separation of the tephras can be displayed graphically on discriminant function axes. A stepwise procedure can be used to identify the most effective discriminating oxides within the dataset. The second part of the analysis applies these discriminant functions to a further geochemical dataset from unknown tephras and classifies these tephras into one of the known groups with a known probability of misclassification. The technique thus provides an objective technique which uses the full complement of geochemical data and yields a measure of similarity between tephras and a probability of identification.

The data assembled for the purposes of this study are listed in Table 1. The table shows that the published data from both Icelandic-type tephras and tephras found in the British Isles are relatively few. Where published data are available, they are often only mean and standard deviations, as in the case of the Svinavatn tephras (Thompson & Bradshaw 1986). As this is a first attempt to assess the potential of the technique, and in order to avoid skewing of the dataset with a large set of values from one or two tephras, only the first ten analyses were used for each reference tephra. In the case of KAL-X and KAL-Y (Dugmore & Newton 1992), a number of different subsets of data were available and ten of each of these were used. The inclusion of the recent Grímsvötn data was to provide additional data from this volcanic system with which to compare the Saksunarvatn ash. Despite the limited data, the range considered here does contain data from those volcanic systems and eruptions most likely to have deposited tephra in Britain and should therefore provide an adequate test for a provisional assessment of the technique. The fact that the data are a product of different microprobe techniques carried out by different operators on different machines may tend to blur differences between tephras, but this is typical of the comparisons which may have to be made and enables further evaluation of the issue of data quality.

One of the issues in comparison of geochemical microprobe data is that of data treatment. Data is usually expressed as raw percentage data, or normalized to 100%

Table 1. *Data sources for DFA analysis*

Reference	Tephra	Volcanic system	n
Reference Icelandic tephras			
Dugmore & Newton (1992)	KAL-X	Hekla 3	30
	KAL-Y	Hekla 4	40
Mangerud et al. (1984)	Saksunarvatn	Grímsvötn	13
Grönvold & Jóhanneson (1984)	Grímsvötn 1983	Grímsvötn	10
	Grímsvötn 1934	Grímsvötn	14
Thompson & Bradshaw (1986)	Svinavatn H1	Hekla 1	10
	Svinavatn H3	Hekla 3	10
	Svinavatn H4	Hekla 4	10
Distal tephras in the British Isles			
Dugmore (1989)	Altnabreac	Hekla 4	10
Bennett et al. (1992)	Dallican 704 cm (Saksunarvatn ash)	Grímsvötn	1
	Dallican 4310	Hekla 4	18
	Dallican 3850	Hekla 4	7
Pilcher et al. (1995)	Sluggan H1	Hekla 1	16
	Sluggan H4	Hekla 4	11

The eruptions for the distal tephras are those suggested by the authors concerned. Svinavatn data was computer generated from the quoted mean and standard deviations in Thompson & Bradshaw (1986). The source of the KAL-X tephra has recently been revised to Hekla-S (Dugmore, pers. comm.). Note: Since the DFA analyses were completed, many more data have become available.

total. Hunt & Hill (1993) suggest the former is more reliable whereas the INQUA guideline recommends the latter (Froggatt 1992). Whichever approach is adopted, the problem of a constant sum remains, which leads to problems in the statistical treatment of the data (Aitchison 1983; Stokes & Lowe 1988). Stokes & Lowe (1988) suggest an alternative approach which is to perform a log-ratio transformation. All three approaches were tried here in order to assess the most effective classification model. The log-ratio data was derived using the following equation:

$$y_i = \log(x_i/x_d)$$

where:

y_i is the log ratio of oxide i, x_i the percentage of oxide i and x_d the percentage of CaO. CaO was chosen as the divisor as it had few zero values, moderate abundance and relatively small variance. The clear disadvantage of this is that the CaO data are not used in the discrimant functions.

DFA of the reference data

Results of the stepwise DFA on the reference dataset are shown in Table 2. The selection of variables was similar for the raw and normalized datasets, with MgO, Al_2O_3 and Na_2O being selected as the first three variables. The order of the other variables is similar, except for SiO_2 which was unimportant in the normalized data. CaO and MgO are excluded from consideration in the log-ratio data as CaO is the divisor and MgO had many zero values. Of the remaining variables, K_2O, SiO_2 and Na_2O are the most important. The discrimant function characteristics are given in Table 3. The three analyses give similar results with over 97% of variance explained

Table 2. *Stepwise DFA of variables significant in distinguishing between tephras*

Step	Raw		Normalized		Log-ratio	
1	MgO	0.01342	MgO	0.01348	K_2O	0.03104
2	Al_2O_3	0.00258	Al_2O_3	0.00336	SiO_2	0.00631
3	Na_2O	0.00077	Na_2O	0.00103	Na_2O	0.00167
4	SiO_2	0.00036	CaO	0.00054	Al_2O_3	0.00109
5	CaO	0.00019	K_2O	0.00031	MnO	0.00078
6	K_2O	0.00011	FeO	0.00023	FeO	*
7	FeO	0.00008	MnO	0.00018	TiO_2	*
	MnO	*	SiO_2	*		
	TiO_2	*	TiO_2	*		

Oxides are listed in the order selected by the stepwise procedure and the value of Wilks' Lambda at each step is given for the three separate analyses.
*Variables that were not selected by the discriminant analysis as being important in distinguishing between tephras in the sample dataset

Table 3. *Characteristics of the discriminant functions in the three analyses*

Function	Eigenvalue	Percentage variance	Canonical correlation
Raw data			
1	84.977	88.26	0.994
2	4.949	5.14	0.912
3	4.299	4.47	0.901
Total		97.87	
Normalized data			
1	86.945	90.16	0.994
2	4.523	4.69	0.905
3	3.962	4.11	0.894
Total		98.95	
Log-ratio data			
1	38.021	83.33	0.987
2	4.180	9.16	0.898
3	2.821	6.18	0.859
4	0.519	1.14	0.585
Total		99.81	

in all datasets. The analyses on raw and normalized percentage data yield results strongly dominated by the first discriminant function. With the log-ratio data, functions 2 and 3 are more important. The separation of the tephras on the first two discriminant functions is shown in Fig. 1. Function 1 clearly separates the two volcanic systems and function 2 provides some separation of the eruptions. The modern Grímsvötn eruptions are inseparable from the Saksunarvatn ash, except with the normalized data, but there are two clear groups of Hekla data in all analyses, separating KAL-X and Svinavatn H3 from the other tephras. The two tephras from the Hekla 4 eruption (Svinavatn H4 and KAL-Y) are inseparable but Svinavatn H1 plots consistently towards the H3–KAL-X group. The classification efficiency is shown in Table 4 and reflects this graphical representation of the results. The major overlaps occur between the two modern Grímsvötn eruptions, the two Hekla 4 tephras and Svinavatn H1. There are also three samples consistently

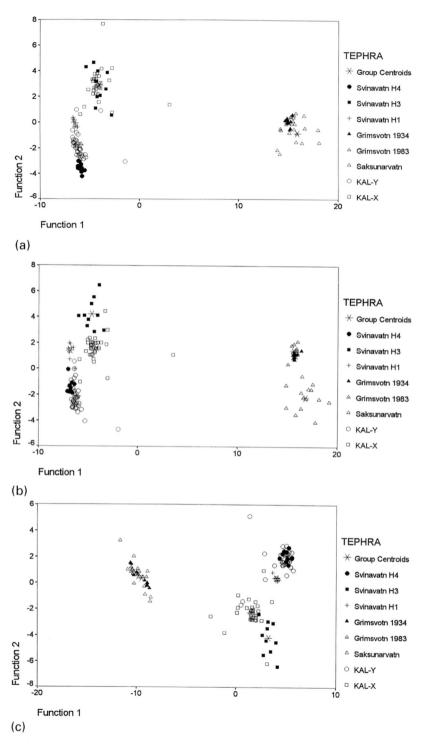

Fig. 1. Plots of the first and second canonical discriminant functions for the eight tephras based on: (**a**) known raw data; (**b**) known normalized; (**c**) known log-ratio data. (a)–(c) include Na_2O data.

Table 4. *Classification efficiency of the reference samples for the raw, normalized and log-ratio datasets*

Actual tephra	n*	\multicolumn{8}{c}{Predicted tephra}							
		1	2	3	4	5	6	7	8
Raw data (90.3%)									
1. KAL-X	30	29	1	–	–	–	–	–	–
2. KAL-Y	40	2	35	–	–	–	3	–	–
3. Saksunarvatn	14	–	–	14	–	–	–	–	–
4. Grímsvötn 1983	10	–	–	–	8	2	–	–	–
5. Grímsvötn 1934	10	–	–	–	4	6	–	–	–
6. Svinavatn H1	10	–	1	–	–	–	9	–	–
7. Svinavatn H3	10	–	–	–	–	–	–	10	–
8. Svinavatn H4	10	–	–	–	–	–	–	–	10
Normalized data (85.1%)									
1. KAL-X	30	29	1	–	–	–	–	–	–
2. KAL-Y	40	2	30	–	–	–	3	–	5
3. Saksunarvatn	14	–	–	14	–	–	–	–	–
4. Grímsvötn 1983	10	–	–	–	8	2	–	–	–
5. Grímsvötn 1934	10	–	–	–	4	6	–	–	–
6. Svinavatn H1	10	–	–	–	–	–	9	–	1
7. Svinavatn H3	10	–	–	–	–	–	–	10	–
8. Svinavatn H4	10	–	1	–	–	–	1	–	8
Log-ratio data (86.2%)									
1. KAL-X	30	28	1	–	–	–	–	1	–
2. KAL-Y	37	2	30	–	–	–	1	–	4
3. Saksunarvatn	14	–	–	14	–	–	–	–	–
4. Grímsvötn 1983	10	–	–	–	8	2	–	–	–
5. Grímsvötn 1934	10	–	–	–	4	6	–	–	–
6. Svinavatn H1	9	–	–	–	–	–	8	–	1
7. Svinavatn H3	10	–	–	–	–	–	–	10	–
8. Svinavatn H4	10	–	1	–	–	–	1	–	8

The overall efficiency is given in parenthesis.
* Reduced in log-ratio data due to zero values.

misplaced between KAL-X and KAL-Y. In terms of overall performance, the raw data analysis was apparently better (90.3% efficiency overall). However, it would be misleading to use this as an indication as several tephras should be the same (H4–KAL-Y) or very similar (Grímsvötn 1934 and 1983) in composition. Therefore, there is little difference between performance of the different data treatments given this rather crude trial dataset.

Treatment of missing data and the sodium problem

The problem of different geochemical analysis techniques was raised above. This has several different facets, but one of the key concerns is over the analysis and use of sodium data. Due to volatilization and electron-induced migration, sodium can be lost when long counting times are used (Neilsen & Sigurdsson 1981; Froggatt 1992; Hunt & Hill 1993). Therefore, some authors do not analyse or exclude Na_2O in analyses (Bennett *et al.* 1992). In this study, sodium data was not available from the Svinavatn tephras (Thompson & Bradshaw 1986) and the Saksunarvatn ash (Mangerud *et al.* 1984). Given this mixed dataset, it is difficult to know whether sodium data can be of any use at all in the discrimination between tephras. All of the

Table 5. *Percentage variance explained by functions 1 and 2 on different datasets including and excluding Na_2O data*

	All data (%)	Excluding Na (%)
Raw data	93.40 (90.30)	96.48 (88.81)
Normalized	94.84 (85.07)	95.17 (84.33)
Log-ratio	92.49 (86.15)	97.82 (77.69)

Values in parentheses are the percentages correctly classified.

analyses described so far include Na_2O, with missing values being replaced by the mean. This means that the performance of the different analyses is difficult to interpret properly as some of the differences between tephras will arise as a result of this missing data. Some idea of the influence of the inclusion of the sodium data can be gained from a comparison of analyses with and without this data.

Table 5 shows a comparison of the explanatory power of the first two functions together with the classification efficiency for the three sets of analyses with and without sodium data included. In general, there are only minor differences between these analyses, although the percentage variance explained is higher when the sodium data are excluded, the classification efficiency is reduced. Seen graphically (Fig. 2), the distribution of tephras is very similar to that where sodium data are included (Fig. 1). Comparing the two raw data plots, there is some reordering of the Hekla tephras along discriminant function axis 2, with a slightly clearer overlap between Svinavatn H3 and KAL-X. The normalized and log-ratio data shows a more robust data structure and the plot is little altered by the exclusion of sodium. The reduction in classification efficiency when sodium data are excluded can be explained by differences in geochemical analysis techniques and data processing. The differences picked out when sodium data are included may be a result of analytical differences rather than real differences in tephra geochemistry. This suggests that where mixed datasets are analysed, it is preferable to exclude elements where there is missing data altogether, otherwise spurious discriminations may be made. Within this small reference dataset, it makes little difference to the discrimination between tephras. The log-ratio data performs best in terms of clear separation of groups of tephras, particularly in discriminating between the two groups of Hekla tephras.

Separation of Hekla eruptives

The stepwise procedure tends to select variables that separate groups which are far apart initially and may not separate groups which are close together so effectively. However, a second analysis of closely related groups can yield improved discriminant functions (Stokes & Lowe 1988). A second analysis was therefore carried out to examine the possibility of improved separation of the Hekla eruptives. The results of the analysis of this more limited dataset are shown in Table 6 for the raw dataset. In terms of the classification efficiency, the results are identical to those obtained from the analysis of the full dataset including the Grímsvötn eruptions. The graphical separation of the tephras is clearer, however, with the only real overlap between Svinavatn H1 and KAL-Y. However, since the separation seems to

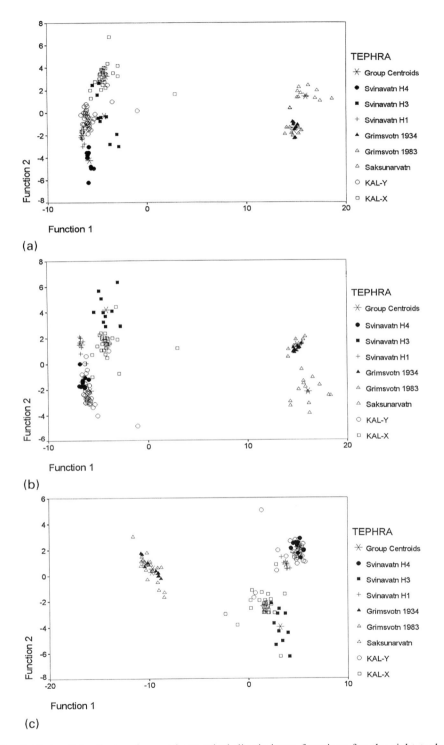

Fig. 2. Plots of the first and second canonical discriminant functions for the eight tephras based on: (**a**) known raw data; (**b**) known normalized data; (**c**) known log-ratio data. (a)–(c) Exclude Na_2O data.

Table 6. *Classification efficiency of partial analysis on Hekla data only compared to that derived from the full dataset*

Actual tephra	n	Predicted tephra				
		1	2	6	7	8
Hekla classification from Hekla only data						
1. KAL-X	30	29	1	–	–	–
2. KAL-Y	40	2	35	3	–	–
6. Svinavatn H1	10	–	1	9	–	–
7. Svinavatn H3	10	–	–	–	10	–
8. Svinavatn H4	10	–	–	–	–	10
Hekla classification from full analysis						
1. KAL-X	30	29	1	–	–	–
2. KAL-Y	40	2	35	3	–	–
6. Svinavatn H1	10	–	1	9	–	–
7. Svinavatn H3	10	–	–	–	10	–
8. Svinavatn H4	10	–	–	–	–	10

Table 7. *Classification of distal tephras found in Britain and Ireland based on DFA of reference dataset including Na_2O*

Tephra Saksun	Svin. H1	Svin. H3	KAL-X	KAL-Y	Svin. H4	G-1934	G-1983
Altnabreac				10			
Dallican 704 cm							1
Dallican 4310 years BP		2	16	1			
Dallican 3850 years BP	1	3	2				
Sluggan H1		1	10	5			
Sluggan H4				11			

be based on analytical differences rather than real differences (H4 and KAL-Y are the same tephra), such an apparently improved separation should be regarded with caution. While this further analysis has not improved the discrimination between the tephras included in this study, different results have been obtained by others (Stokes & Lowe 1988; Stokes *et al.* 1992) and a larger dataset may ultimately provide a similar improvement for the Icelandic tephras.

Classification of distal tephras from British deposits

The ultimate aim of DFA on the Icelandic tephras would be to enable classification of the distal tephras found in Britain and Ireland. The classification in Table 7 is based on the DFA of the raw dataset, excluding Na_2O. Several of these tephras have been identified as Hekla 4 on the basis of their radiocarbon dates and geochemistry. In this analysis, Altnabreac (Dugmore 1989) and Sluggan H4 (Pilcher *et al.* 1995) fall clearly into KAL-Y, but the two mid-Holocene tephras from Dallican Water (Bennett *et al.* 1992) appear to be allied more closely with Svinavatn H3 and KAL-X on the basis of this analysis. Bennett *et al.* (1992) are rightly cautious in the discussion of the identity of these tephras and do not claim a definite identity but

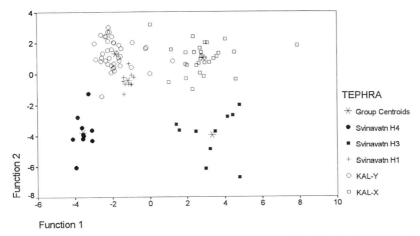

Fig. 3. Plot of the first and second canonical discriminant functions for the five Hekla-derived tephras based on raw data, including Na_2O data.

suggest Hekla 4 is the most likely origin on age grounds. However, while the data used for the DFA are relatively crude, the results demonstrate clearly the dangers of assigning identity on the basis of age grounds. It appears quite possible that the tephras from 4310 and 3850 years BP at Dallican may be derived from the same eruption as KAL-X, which is only stratigraphically correlated with another Hekla eruption from c. 3500 years BP (Dugmore & Newton 1992), particularly considering the potential errors in radiocarbon dating and stratigraphic correlation between the Icelandic sites. This is exactly the sort of difficulty which is likely to arise more frequently in the future with the discovery of further closely spaced distal tephras.

Conclusion: problems and potential of DFA and British tephras

The analyses presented here were carried out on a relatively small selection of data from only a few of the larger Holocene eruptions in Iceland which are potential sources for distal tephras found in Britain and Ireland. In addition, the data was gathered from disparate sources with inconsistent analytical techniques. The results should therefore not be regarded as definitive in any sense and further analyses will be carried out on an improved dataset in future. In particular, besides increasing the quantity and quality of the data used in the analysis, further attention should be given to the evaluation of data pretreatment. Stokes & Lowe (1988) suggest that log-ratio data should be used in conjunction with a principal components analysis to remove outlying data points for the best discriminant functions and this technique should be assessed in the British–Icelandic context. However, the analyses do allow a provisional assessment of the utility of such an approach for the improved identification and correlation of tephras in the future:

- the separation of the two volcanic centres is very clear, as would be expected from a comparison of the mean and standard deviation figures for geochemistry. The Hekla material is all highly silicic compared to the basaltic Grímsvötn material. The inclusion of further volcanic centres will be needed to

assess this further, perhaps especially of the Oraefajökull and Katla systems;
- the separation of eruptions within individual volcanic systems is more variable. Grímsvötn 1983 and 1934 are indistinguishable and often overlap with the Saksunarvatn ash, depending on which analysis is carried out. The Hekla eruptions break down into two main groups; Svinavatn H3–KAL-X and H1–H4–KAL-Y. There is some improvement in this separation when a partial analysis is carried out but, given the current dataset, differences in analytical procedures cannot be separated from real differences between tephras;
- a comparison of data pretreatment suggests that different procedures make only minor differences to the resulting discriminant functions. In addition, where missing values are present, it may be better to exclude the variable from consideration in the analysis, rather than include it and accept spurious tephra separations. However, the relatively minor effects of data pretreatment and inclusion or exclusion of oxides suggests that DFA is a robust technique even when comparing mixed datasets.
- it is apparent from the analysis of this limited dataset, that the classification of tephras located in British sediments from a consideration of major oxides content alone, may not be adequate to discriminate between all eruptions. Consideration should be given to the adoption of analytical techniques which provide geochemical data on the rare earth content of tephra shards, which has been demonstrated to discriminate between tephras which are otherwise very similar (Shane & Froggatt 1994). This technique would complement, rather than replace, the major oxide data;
- the way forward with the application of DFA to the identification of distal tephras in Britain and Ireland lies with the establishment and analysis of a large and coherent reference dataset. This has advanced considerably since the analyses presented here were carried out and a large quantity of data are now available through Tephrabase, a worldwide-web database at the University of Edinburgh (*http://www.geo.ed.ac.uk/tephra/tbasehom.html*). The analysis of these and other data, and the provision of discriminant functions based on them, should allow a more objective assessment of tephra identity even in the absence of radiocarbon dates. This is essential if Icelandic tephras are to be widely used as isochrones in peat and lake deposits in Britain and Ireland.

References

AITCHISON, J. 1983. Principal components analysis of compositional data. *Biometrika*, **70**, 57–65.
BEGET, J., MASON, O. & ANDERSON, P. 1992. Age, extent and climatic significance of the c. 3400 BP Aniakchak Tephra, western Alaska. USA. *The Holocene*, **2**, 51–56.
BENNETT, K. D. 1994. Tephra geochemistry: a comment on Hunt and Hill. *The Holocene*, **4**, 435–438.
——, BOREHAM, S., SHARP, M. J. & SWITSUR, V. R. 1992. Holocene history of environment, vegetation and human settlement of Catta Ness, Lunnasting, Shetland. *Journal of Ecology*, **80**, 241–273.
BLACKFORD, J. J., EDWARDS, K. J., DUGMORE, A. J., COOK, G. T. & BUCKLAND, P. 1992. Hekla-4: Icelandic volcanic ash and the mid-holocene Scots Pine decline in northern Scotland. *The Holocene*, **2**, 260–265.
BORCHARDT, G. A., ARUSCAVAGE, P. J. & MILLARD, H. T. 1972. Correlation of the Bishop Ash, a Pleistocene marker bed, using instrumental neutron activation analysis. *Journal of*

Sedimentary Petrology, **42**, 301–306.
BRAITSEVA, O. A. & MELEKETSEV, I. V. 1990. Eruptive history of Karymsky volcano, Kamchatka, USSR, based on tephra stratigraphy and ^{14}C dating. *Bulletin Volcanologique*, **53**, 195–206.
BUCKLAND, P. C., FOSTER, P., PERRY, D. W. & SAVORY, D. 1981. Tephrochronology and palaeoecology: the value of isochrones. *In:* SELF, S. & SPARKS, R. S. J. (eds) *Tephra studies*, 381–389. Reidel, Dordrecht.
CHARMAN, D. J. GRATTAN, J. P. & WEST, S. 1995. Environmental response to tephra deposition in the Strath of Kildonan, northern Scotland. *The Journal of Archaeological Science*, **22**, 799–809.
DUGMORE, A. J. 1989. Icelandic volcanic ash in Scotland. *Scottish Geographical Magazine*, **105**, 168–172.
——, BUCKLAND, P. 1991. Tephrochronology and Late Holocene Soil Erosion in South Iceland. *In:* MAIZELS, J. K. & CASELDINE, C. (eds) *Environmental Change in Iceland: Past and Present*, Kluwer, 147–159.
——, NEWTON, A. J. 1992. Thin tephra layers in Peat revealed by X-radiography. *Journal of Archaeological Science*, **19**, 163–170.
FROGGATT, P. C. 1992. Standardisation of the chemical analysis of tephra deposits. Report of the ICCT working Group. *Quaternary International*, **13/14**, 93–96.
GRAHAM, D. K., HARLAND, R., GREGORY, D. M, LONG, D. & MORTON, A. C. 1990. The biostratigraphy and chronostratigraphy of BGS Borehole 78/4, North Minch. *The Scottish Journal of Geology*, **26**, 65–75.
GRONVOLD, K. & JOHANNESSON, H. 1984. Eruption in Grímsvötn, 1983. *Jökull*, **34**, 1–11.
HALL, V. A., PILCHER, J. R. & MCCORMAC, F. G. 1993. Tephra dated lowland landscape history of the north of Ireland, A.D. 750–1150. *New Phytologist*, **125**, 193–202.
——, —— & —— 1994. Icelandic volcanic ash and the mid-Holocene Scots pine (*Pinus sylvestris*) decline in the north of Ireland: no correlation. *The Holocene*, **4**, 79–83.
HUNT, J. & HILL, P. 1993. Tephra geochemistry: a discussion of some persistent analytical problems. *The Holocene*, **3**, 271–278
—— & —— 1994. Geochemical data in tephrochronology: a reply to Bennett. *The Holocene*, **4**, 436–438.
KELLER, J. 1981. Quaternary teprochronology in Mediterranean regions. *In:* SELF, S. & SPARKS, R. S. J. (eds) *Tephra Studies*, Reidel, 227–244.
KYLE, P. R., JEZEK, P. A., MOSLEY-THOMPSON, E. & THOMPSON L. G. 1981. Tephra layers in the Byrd station ice core and the Dome C ice core, Antarctica and their climatic significance. *Journal of Volcanology and Geothermal Research*, **11**, 29–39.
LARSEN, G. & THÓRARINNSON, S. 1977. H^4 and other acid Hekla tephra layers. *Jökull*, **27**, 28–49.
LONG, D. C. & MORTON, A. C. 1987. An ash fall within the Loch Lomond Stadial. *Journal of Quaternary Science*, **2**, 97–101.
LOWE, D. J. 1988. Late Quaternary Volcanism in New Zealand: towards an integrated record using distal airfall tephras in lakes and bogs. *Journal of Quaternary Science*, **3**, 111–120.
MANGERUD, J., LIE, S. L., FURNES, H., KRISTIANSEN, I. L. & LOMO, L. 1984. A younger Dryas ash bed in western Norway, and its possible correlations with Tephra cores from the Norwegian Sea and the North Atlantic. *Quaternary Research*, **21**, 85–104.
NIELSEN, C. H. & SIGURDSSON, H. 1981. Quantitative methods for electron microprobe analysis of sodium in natural and synthetic glasses, *American Mineralogist*, **66**, 547–52.
PERSSON, C. 1971. Tephrochronological investigations of peat deposits in Scandinavia and on the Faroe islands. *Sveriges Geologiska Undersköning*, **Series C**, 656.
PILCHER, J. R. & HALL, V. A. 1992. Towards a tephrochronology for the north of Ireland. *The Holocene*, **2**, 255–259.
——, —— & MCCORMAC, F. G. 1995. Dates of Holocene icelandic volcanic eruptions from tephra layers in Icelandic peats. *The Holocene*, **5**, 103–110.
——, BAILLIE, M. G. L., SCHMIDT, B. & BECKER, B. 1984. A 7272-year tree ring chronology for western europe. *Nature*, **312**, 150–152.
RABEK, K., LEDBETTER, M. T. & WILLIAMS, D. F. 1985. Tephrochronology of the western gulf of Mexico for the last 185 000 years. *Quaternary Research*, **23**, 403–416.

Rose, W. I. & Chesner, C. A. 1990. Worldwide dispersal of gas and ashes from the Earth's largest known eruption: Toba, Sumatra, 75 ka. *Palaeogeography, Palaeoclimatology, Palaeoecology*, **89**, 269–275.

Shane, P. A. R. & Froggatt, P. C. 1994. Discriminant function analysis of New Zealand and North American tephra deposits. *Quaternary Research*, **41**, 70–81.

St Seymour, K. & Christianis, K. 1995. Correlation of a tephra layer in Western Greece with a late Pleistocene eruption in the Campanian Province of Italy. *Quaternary Research*, **43**, 46–54.

Stanley, D. J. & Sheng, H. 1986. Volcanic shards from Santorini (Upper Minoan Ash) in the Nile Delta. *Nature*, **320**, 733–35.

Stokes, S. & Lowe, D. J. 1988. Discriminant function analysis of late Quaternary Tephras from five volcanoes in New Zealand using glass shard major element chemistry. *Quaternary Research*, **30**, 270–283.

——, —— & Froggatt, P. C. 1992. Discriminant function analysis and correlation of late Quaternary tephra deposits from Taupo and Okataina volcanoes, New Zealand, using glass shard major element compositions. *Quaternary International*, **13/14**, 103–120.

Sullivan, D. G. 1988. The discovery of Santorini Minoan tephra in western Turkey. *Nature*, **333**, 552–554.

Thompson, W. & Bradshaw, R. H. 1986. The distribution of ash in Icelandic lake sediments and the relative importance of mixing and erosion processes. *Journal of Quaternary Science*, **1**, 3–11.

Thórarinsson, S. 1967. *The eruption of Hekla 1947–1948. The eruptions of Hekla in Historical Times.* H.F. Leiftur.

—— 1980. Langleithir, gjósku úr Thremur Kötlugosum. *Jökull*, **30**, 65–73.

Regional warming of the lower atmosphere in the wake of volcanic eruptions: the role of the Laki fissure eruption in the hot summer of 1783

JOHN GRATTAN[1] & JON SADLER[2]

[1] *The University of Wales, Institute of Geography and Earth Sciences, Aberystwyth SY23 3DB, UK*
[2] *The University of Birmingham, School of Geography, Birmingham B15 2TT, UK*

Abstract: A suggestion is made that the gases emitted in the Laki fissure eruption in the summer of 1783 contributed to the high surface-air temperatures recorded in many parts of Europe. This paper presents European documentary and instrumental evidence which suggests that, during the initial stages of the eruption of the Laki fissure in June and July 1783, intense localized warming of the lower atmosphere occurred in many parts of Europe. This phenomenon is examined in detail and a hypothesis advanced that, under specifically defined conditions, volcanic gases may lead to warming of the lower atmosphere.

It has been recognized that the initial phases of the Laki fissure eruption coincided with the high surface-air temperatures recorded in Europe during the summer of 1783 (Wood 1984, 1992), but no model has yet been advanced which may adequately account for this phenomenon. Observations of weather and temperature records made in Britain in the summer of 1783 (Manley 1974; Kington 1980) are contrary to established models of climatic response to Northern Hemisphere eruptions (Lamb 1970; Kelly & Sear 1984; Sear *et al.* 1987). There is little evidence for either rapid surface cooling or unstable circulation, in fact, the reverse appears to be true. Qualitative and quantitative data from Europe suggests that late June and July in particular were notably hot. July 1783 was the hottest recorded in Manley's (1974) collation of central England temperatures, and also in Copenhagen (Kington, 1980). Elsewhere in Europe, July temperatures were above average in Stockholm, Berlin and Geneva (Wood 1992). Atmospheric circulation patterns described by Kington (1980, 1988) establish the presence of a relatively stable, high-pressure air mass over Europe from June 22nd to July 20th, a period which coincides with the peak productivity of the Laki fissure eruption (Thórarinsson 1969; Sigurdsson 1982; Metrich *et al.* 1991; Thordarson & Self 1993), and with descriptions of a hot dry fog and records of high surface-air temperatures (detailed below). An analysis of the atmospheric and climatic phenomena recorded during the summer of 1783 illustrates the clear relationship between volcanic gases and high surface-air temperatures at this time.

The possibility of short-term tropospheric warming in the wake of a volcanic eruption has been considered a theoretical possibility, with the limited potential to counterbalance aerosol-induced cooling in the first months after an eruption

(Baldwin *et al.* 1976; Pollack *et al.* 1976). However, investigations of the potential of volcanic eruptions to affect climate have concentrated on the cooling effect which follows the introduction of fine particulate matter and acid gases into the stratosphere, thereby increasing the global albedo (Hansen *et al.* 1978; Harshvardhan 1979; Chou *et al.* 1984). Discussions of climatic response to volcanic eruptions have emphasized this aspect on a hemispheric scale (Kelly 1977; Kelly & Sear 1984; Lamb 1988; Luhr 1991). Currently, estimates of the degree to which volcanic eruptions may affect climate, and the assumptions made based on these, have been subject to a considerable degree of review. Many studies have progressed beyond attempting to correlate volcanic eruptions with relatively minor and ephemeral temperature responses and atmospheric circulation disruption (Sigurdsson 1982; Rampino *et al.* 1985; Rampino 1988; Handler & Andsager 1990), and focused instead on local and regional impacts. These have suggested that not all resultant climatic disruption need result in a reduction of surface-air temperatures and an intensification of environmental stress and, indeed, that the reverse may be true (Groisman 1992; Robock & Mao 1992; Cleaveland 1992). Many models have emphasized the need for a clear injection of volcanic material into the stratosphere as a prerequisite for subsequent surface cooling of hemispheric temperatures (Baldwin *et al.* 1976; Pollack *et al.* 1976; Sigurdsson, 1990). Fissure eruptions may not achieve a clear injection of material into the stratosphere, and the majority of gas and particulate material emitted may be confined to the troposphere (Tripoli & Thompson 1988). It is therefore unreasonable to expect these models, devised to account for the behaviour of explosive eruptions which penetrate the stratosphere, to accurately describe the climatic impact of a fissure eruption and the emission of vast quantities of sulphur gases into the troposphere. A study of the climatic phenomena associated with the Laki fissure eruption may help to develop these models.

The summer of 1783: Qualitative data

Reliable contemporary observers have left records which establish the presence of considerable levels of volcanogenic material in the lower atmosphere in the summer of 1783, similar accounts may be found after many historic eruptions (Stothers & Rampino 1983). Most of the observations made in the summer of 1783, however, are unique in their association with hot rather than cool weather. White (1789), a careful observer of natural phenomena, described both the atmospheric phenomena and the summer of 1783 in great detail:

> The summer of 1783 was an amazing and portentous one, and full of horrible phenomena; for besides the alarming meteors and thunder-storms that affrighted many counties of this kingdom, the peculiar haze or smokey fog, that prevailed for many weeks in this island and in every part of Europe, and even beyond its limits, was a most extraordinary appearance, unlike anything known within the memory of man. By my journal I find that I had noticed this strange occurrence from June 23 to July 20 inclusive, during which period the wind varied to every quarter without making any alteration in the air. The sun, at noon, looked as blank as a clouded moon, and shed a rust coloured ferruginous light on the ground, and floors of rooms; but was particularly lurid and blood coloured at rising and setting. All the time the heat was so intense that butchers meat could hardly be eaten on the day after it was killed; and the flies swarmed

so in the lanes and hedges that they rendered the horses half frantic, and riding irksome.

On June 25th Parson Woodforde, at Weston Longeville Norfolk, recorded in his diary (1984):

Very uncommon lazy and hot weather. The sun very red at setting and on July 28th:

This has been the hottest day this year, and I believe the hottest that I ever felt, many say the same.

Other material presents similar evidence. The meteorological register kept at Lyndon Hall, Rutland, by Thomas Barker (unpublished) mentions 'very thick smoaky air' on June 25th. On June 26th he recorded very hot temperatures and an obscured sun:

calm, very thick air, very hot till evening, sun scarce shone.

Hot blue air, and thick smokey air are regularly mentioned in his weather journal from June 22nd to July 21st and he summarized July as follows:

Two thirds of July was the same thick blue air as June ended with the month mainly hot, sometimes very hot.

The difference in the atmosphere after July 21st was so remarkable that on July 25th he noted that it was:

...warm and pleasant, thickness of the air gone off.

Many correspondents confirmed Barker's observations and clearly associate hot temperatures, mists and a red sun. William Dunn (unpublished), a schoolmaster of Belbroughton, Worcestershire, described much of July as 'very hot!' and gave a graphic description of the state of the atmosphere on July 10th:

The air continues in a putrid state, the sun is as red as blood.

Horace Walpole, writing on July 15th (Cunningham 1938), also appeared to link volcanogenic aerosols with unseasonable heat and a constant mist:

I am sorry your Ladyship has suffered so much by the heat...I am tired of this weather...it parches the leaves, makes the turf crisp...and keeps one in a constant mist that gives no dew but might as well be smoke. The sun sets like a pewter plate red hot.

Newspapers also commented on the high temperatures and the state of the air. *Parker's General Advertiser* and *Morning Intelligence* recorded on July 11th:

Perhaps since England's origin there has not been a tract of weather so warm and hot as from the 23rd of June to the 9th of July.

Nor were such descriptions limited to the British Isles. In contrast to the inference drawn by Franklin (1784), who was in Europe in 1783, that the summer was cooler than normal. Similar comments were found in the letters columns of serious newspapers, a 'letter from Paris' noted that:

For a considerable time past, the weather has been very remarkable here; a kind of hot fog obscures the atmosphere, and gives the sun much of that dull red appearance which wintry fogs sometimes produce. The fog is not peculiar to Paris, those who come lately from Rome say it is as thick and hot in Italy, and even the top of the Alps is covered

with it, and travellers from Spain affirm the same of that kingdom.
General Evening Post, 12th July 1783; *The Ipswich Journal*, 19 July 1783.

Extensive material reviewing the wider European picture is presented in Brayshay & Grattan (1999) and in Stothers (1996).

Contemporary newspaper accounts also associate a sulphurous stench and the intense heat:

> where there was no rain the sulphurous stench remained in the air ... when the heat was more intense than before.
> *Bristol Journal*, 19th July 1783.

All the accounts above suggest that the atmospheric phenomena, the 'peculiar haze, or smokey fog', the obscured noonday sun, the lurid colours at sunrise and sunset, and reduced visibility, may be the result of the concentration of volcanic gases and aerosols in the atmosphere. The reduction of summer visibility has been associated with the long-range transport of pollutants in the lower atmosphere (Leavey & Sweeney 1990). Descriptions of severe acid damage to vegetation can also be found (Grattan & Charman 1994; Grattan & Gilbertson 1994; Grattan & Pyatt 1994). The following by Cullum (1783) is typical of many reports:

> The aristæ of the barley ... became brown and weathered at their extremities, as did the leaves of the oats; the rye had the appearance of being mildewed ... The Larch, Weymouth Pine, and hardy Scotch fir, had the tips of their leaves withered; ... Cherry-trees, a standard peach tree, filbert and hasel-nut-trees, shed their leaves plentifully, and littered the walks as in autumn. ... All these vegetables appeared exactly as if a fire had been lighted near them, that had shrivelled and discoloured their leaves.

Taken together, all these accounts clearly establish the presence of significant levels of sulphur gases and sulphuric acid aerosols in the lower atmosphere and the most likely source for these high concentrations is the Laki fissure. In all the above accounts descriptions of sulphurous stench, dry haze, blue air, thick smokey air and the red appearance of the sun at rising and setting are all linked to descriptions of intense heat. Quite clearly, the meteorological conditions outlined by Kington (1988) – the stable, high-pressure air mass – contributed to the concentration of gases and the formation of aerosols in the lower atmosphere. Similar events are increasingly common today where anthropogenic sources provide the gases and particulate material, but are rare in an eighteenth century context (Bernardini *et al.* 1987).

The summer of 1783: Quantitative data

Quantitative studies confirm the qualitative descriptions of contemporary observers. In the central England temperature record (Manley 1974), July temperatures for 1783 were exceptional and were not surpassed for 200 years. In contrast, both June and August lie close to the mean for 1770–1795. The June mean temperature for 1783, (14.8°C), is only 0.14°C above the 26 year mean, while August 1783 (15.8°C), was slightly cooler. The mean temperature for July 1783 (18.8°C), was the warmest in the central England temperature record and 2.7°C above the mean July temperature for 1770–1795 (Fig. 1). This was the highest July temperature on record until 1983 (Parker *et al.* 1992), which illustrates the anomalous nature of the temperatures recorded in 1783. Warmer than average July temperatures were not

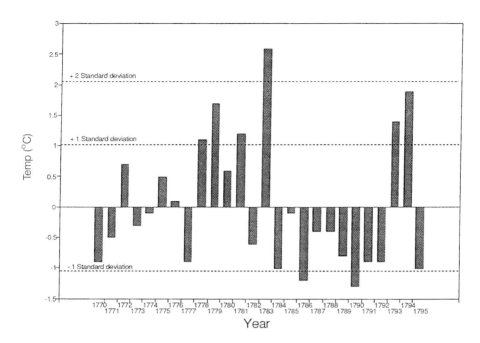

Fig. 1. Variation from July mean temperatures in the central England temperature record.

limited to central England. Mossman's records for Edinburgh (Mossman 1896) shows that July 1783 was the second warmest July in the series from 1764–1896. Similarly, July was the warmest month in the Copenhagen temperature series from 1768–1893 and the fifth warmest in central Europe (Kington 1980).

The monthly mean temperature data does not do justice to the excessive daily temperatures recorded at this time, which were frequently in excess of 24°C and approached 30°C on some occasions. Daily temperature records also serve to confirm the unique nature of the weather in late June and July 1783. Two records of daily temperature considered here, the journal of Thomas Barker (unpublished) and the *London Weather Diary* (unpublished) (both held in the National Meteorological Library and Archive), confirm White's (1789) observation that high temperatures began on June 23rd. Daily records of temperature clearly point to illustrate a sharp increase in temperature on the 22nd June and a decline from July 20th onwards. In Barker's (unpublished) record (Fig. 2) the maximum daily temperatures increased from 14.7°C on June 21st to 23.2°C on June 23rd. The highest temperatures in this record (July 2nd 26.8°C), and July 11th (29.1°C) are accompanied by descriptions of thick air – a common notation in Barker's record between June 23rd and July 20th. The low temperature recorded by Barker on July 16th coincided with the brief positioning of a low-pressure system over the British Isles. *The London Weather Diary* confirms Barker's record (Fig. 3). While this is an admittedly limited dataset, all confirm the descriptions of hot temperatures between June 23rd and July 20th associated with a dry fog and other 'amazing' phenomena.

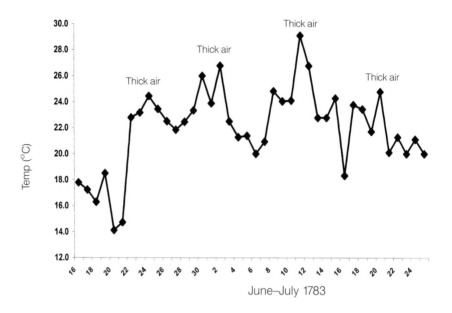

Fig. 2. Daily temperature record, after Thomas Barker (unpublished).

Synthesis: quantitative and qualitative data

The phenomena described above, i.e. the high surface-air temperatures, the acid damage to plants, tremendous thunderstorms and reduced visibility, coincide within a specific period of time between June 23rd and July 20th. This period broadly corresponds with the peak productivity of the Laki fissure (Thordarson & Self 1993) and with the presence over northwest Europe of a relatively stable, high-pressure air cell (Kington 1988). The qualitative data presented above are best explained by the presence of considerable quantities of acid gases in the lower atmosphere, perhaps concentrated by air circulation via the model illustrated in Fig. 4, and therefore both the qualitative and quantitative data clearly links the high temperatures recorded with the presence of significant quantities of volcanic gases in the lower atmosphere.

Discussion

The gases emitted in the early eruptive episodes of the Laki fissure eruption, which were concentrated in the lower atmosphere over western Europe, appear to have either absorbed incoming solar radiation near the ground or inhibited outgoing radiation in excess of the reduction of incoming solar radiation, with the result that net warming occurred. Alternatively, both these models may have operated in tandem with the same net result. The net radiation budget of the Earth is attained through the balance between incoming solar insolation and energy deficits, resulting

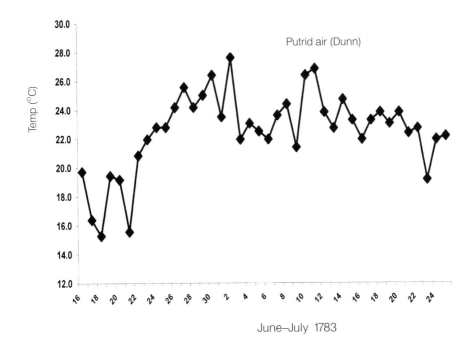

Fig. 3. Daily temperature record, after the *London Weather Diary* (unpublished).

from emitted long-wave and reflected short-wave radiation (Ardanuy et al. 1992). It is unlikely that during June and July 1783 a stratospheric aerosol had formed sufficiently to have increased optical depth and led to cool surface-air temperatures. The photochemical conversion of volcanic gases to an aerosol form within the stratosphere, has been observed to be relatively slow, and maximum perturbation of global albedo may not occur until several months after an eruption when the stratospheric aerosol veil forms (Castleman et al. 1974; Ardanuy et al. 1992). It is therefore clear that any warming mechanism must have operated rapidly in 1783, before the formation of a stratospheric aerosol. Fortunately, this phenomenon can be accurately modelled based on observations of the heating effect of anthropogenic aerosols (Preining 1991, 1992) and the heating effect consequent to the absorption of solar energy by SO_2 gas (Lary et al. 1994).

In observations of the Mount Pinatubo gas cloud, in the period before widespread aerosol formation had occurred, Lary et al. (1994) noted that the SO_2 layer in the atmosphere strongly absorbed solar energy in both the ultraviolet and visible region of the spectrum. As a result, solar energy was deposited within the SO_2 layer and warming occurred. Lary et al. (1994) suggested that this absorption could lead to a short-lived enhancement of air temperature of up to $1\,\text{K}\,\text{day}^{-1}$. Similarly, Preining (1991, 1992), Nkederim (1988) and Wakamatsu et al. (1990) have noted that short-wave absorption by natural and anthropogenic aerosols near the ground causes heating that has been estimated at $1\,\text{K}\,\text{day}^{-1}$.

The studies described above provide the theoretical framework by which it may be

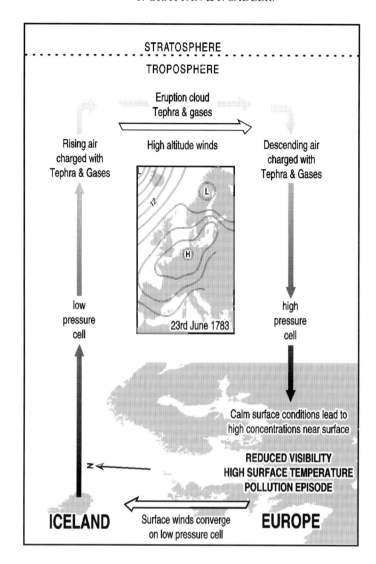

Fig. 4. Air circulation and weather phenomena in Europe, June–July 1783. Synoptic map redrawn from Kington (1988).

suggested that the volcanic gases emitted in the Laki fissure eruption, and concentrated by atmospheric circulation near the ground surface, over a restricted part of Europe, absorbed solar energy and terrestrial radiation and were responsible for the high surface-air temperatures recorded for a brief period in June and July 1783. SO_2 gases may absorb solar energy, resulting in an enhanced diurnal heating rate (Lary et al. 1994), and the subsequent formation of an aerosol from the gas will alternatively generate heating as the result of short-wave absorption. There seems no reason to suggest that SO_2 gases concentrated in the lower atmosphere in 1783 should not have absorbed thermal energy in the manner described by Lary et al.

(1994).

If either, or both, of these models described above, operated in 1783, the subsequent heating effect could have been sufficient to enhance surface-air temperatures to the record-breaking levels recorded in June and July 1783. It must be reiterated that the exceptional phenomena were observed under a particular set of synoptic conditions, i.e. while a high-pressure cell remained over Europe. Temperatures and atmospheric conditions returned to normal once the high-pressure cell had dispersed, since August temperatures were anything but exceptional. It appears that it was the concentration of acid gases by atmospheric air circulation, which directly led to an enhancement of surface-air temperatures (Fig. 4). The subsequent absorption of solar energy by the SO_2 significantly enhanced temperatures, which would normally be the result of a high-pressure cell, positioned over Europe and hence generated the exceptional, and anomalous, temperatures recorded.

The summer of 1783 demonstrates that eruptions, which do not penetrate the stratosphere, have the potential to generate significant warming of the lower atmosphere, a factor not accounted for in current models of climatic response to volcanic eruptions. Recent attention has focused on the potential for conventional explosive eruptions to counterbalance anthropogenically induced greenhouse warming (Hansen & Lacis 1990; Luhr 1992; Kerr 1994). The 'amazing and portentous summer' of 1783 suggests that an alternative short-term scenario is considered, where a concentration of volcanic gases in the near-surface layer of the atmosphere may result in exceptionally warm surface-air temperatures on a regional scale.

This work has been supported by funding from the Scottish Development Department, and grants from SERC and the Leverhulme Trust. Mr R. Castignetti and M. Brayshay assisted in the collation of documentary material. T. Absalom of the University of Plymouth Cartographic Services Unit drew Fig. 4.

References

ARDANUY, P. E., LEE KYLE, H. & HOYT, D. 1992. Global relationships among the Earth's radiation budget, cloudiness, volcanic aerosols and surface temperature. *Journal of Climate*, **5**, 1120–1139.

BALDWIN, P., POLLACK, J. B., SUMMERS, A., TOON, O. B., SAGAN, C. & VAN CAMP, W. 1976. Stratospheric Aerosols and climate change. *Nature*, **263**, 551–555.

BARKER, T. 1777–1789. *Meteorological register made at Lyndon Hall, Rutland*. Unpublished Manuscript: available in the National Meteorological Library and Archive, Bracknell.

BERNARDI, A., CAMUFFO, D., DEL TURCO, A., GAIDANO, D. & LAVAGNINI, I. 1987. Pollution episodes in Venice related to weather types: an analysis for a better predictability. *Science Of The Total Environment*, **63**, 259–297.

BRAYSHAY, M. & GRATTAN, J. P. 1999. Environmental and social responses in Europe to the 1783 eruption of the Laki fissure volcano in Iceland: a consideration of contemporary documentary evidence. *This volume*.

CASTLEMAN, A. W., MUNKELWITZ, H. R. & MANOWITZ, B. 1974. Isotopic studies of the sulfur component of the stratospheric aerosol layer. *Tellus*, **26**, 222–234.

CHOU, M.-D., PENG, L. & ARKING, A. 1984. Climate studies with a multi-layer balance model. Part III: Climatic impact of stratospheric volcanic aerosols. *The Journal of Atmospheric Science*, **41**, 759–767.

CLEAVELAND, M. K. 1992. Volcanic effects on Colorado plateau douglas fir-tree rings. *In:*

HARINGTON, C. R. (ed.) *The Year Without a Summer? World Climate in 1816*. Canadian Museum of Nature, 115–123.
CULLUM, REV. SIR JOHN. 1783. Of a remarkable frost on the 23rd of June, 1783. *Philosophical Transactions of the Royal Society of London, Abridged Volume 1781–1785*, **15**, 604.
CUNNINGHAM, P. 1938. *The Letters of Horace Walpole. Volume VIII*. Richard Bentley.
DUNN, W. *Diary with some accounts, of William Dunn, Schoolmaster of Belbroughton Worcs., 1777–95, including daily reports of the weather*. Unpublished manuscript in the Bodleian Library. Ms. Don. C.76.
FRANKLIN, B. 1784. Meteorological imaginations and conjectures. *Memoirs of the Literary and Philosophical Society of Manchester*, **2**, 373–377.
GRATTAN, J. P. & CHARMAN, D. J. 1994. The palaeoenvironmental implications of documentary evidence for impacts of volcanic volatiles in western Europe, 1783. *The Holocene*, **4**, 101–106.
—— & GILBERTSON. D. D. 1994. Acid-loading from Icelandic tephra falling on acidified ecosystems as a key to understanding archaeological and environmental stress in northern and western Britain. *The Journal of Archaeological Science*, **21**, 851–859.
—— & PYATT, F. B. 1994. Acid damage in Europe caused by the Laki Fissure eruption – an historical review. *The Science of the Total Environment*, **151**, 241–247.
GROISMAN, P. YA. 1992. Possible regional consequences of the Pinatubo eruption: an empirical approach. *Geophysical Research Letters*, **19**, 1603–1606.
HANDLER, P. & ANDSAGER, K. 1990. Volcanic aerosols, El Niño and the southern oscillation. *International Journal of Climatology*, **10**, 413–424.
HANSEN, J. E. & LACIS, A. A. 1990. Sun and dust versus greenhouse gases: an assessment of their relative roles in global climatic change. *Nature*, **346**, 713–719.
—— WANG, W. & LACIS A. A. 1978. Mt. Agung eruption provides test of global climatic perturbation. *Science*, **199**, 1065–1068.
HARSHVARDHAN, H. 1979. Perturbation of the zonal radiation balance by a stratospheric aerosol layer. *Journal of Atmospheric Science*, **36**, 1274–1285.
KELLY, P. M. 1977. Volcanic dust veils and north Atlantic climate change. *Nature*, **268**, 616–617.
—— & SEAR, C. B. 1984. Climatic impact of explosive volcanic eruptions. *Nature*, **311**, 740–743.
KERR, A. 1994. Did Pinatubo send climate warming gases into a dither. *Science*, **263**, 1562.
KING, J. & RYSKAMP, C. 1981. *The Letters and Prose Writings of William Cowper. Volume II*. Clarendon Press.
KINGTON, J. A. 1980. July 1783: The warmest month in the Central England temperature series. *Climate Monitor*, **93**, 69–73.
—— 1988. *The Weather for the 1780s Over Europe*. Cambridge University Press.
LAMB, H. H. 1970. Volcanic dust in the atmosphere; with an assessment of its meteorological significance. *Philosphical Transactions of the Royal Society of London*, **A266**, 425–533
—— 1988. Volcanoes and climate: an updated assessment. In: LAMB, H. H (ed.) *Weather, Climate and Human Affairs*. Routledge, 301–328.
LARY, D. J., BALLUCH, M. & BEKKI, S. 1994. Solar Heating rates after a volcanic eruption: the importance of SO_2 absorption. *Quarterly Journal of the Royal Meteorological Society*, **120**, 1683–1688.
LEAVEY, M. & SWEENEY, J. 1990. The influence of long range transport of air pollutants on summer visibility at Dublin. *International Journal of Climatology*, **102**, 191–201.
LONDON WEATHER DIARY. Unpublished manuscript: available in the National Meteorological Library and Archive, Bracknell.
LUHR, J. F. 1991. Volcanic shade causes cooling. *Nature*, **354**, 104–105.
MANLEY, G. 1974 Central England temperatures: monthly means 1659–1973. *Quarterly Journal of the Royal Meteorological Society*, **100**, 389–405.
METRICH, N., SIGURDSSON, H., MEYER, P. M. & DEVINE, J. D. 1991. The 1783 Lakagigar eruption in Iceland: geochemistry, CO_2 and sulfur degassing. *Contributions to Mineralogy and Petrology*, **107**, 435–477.
MOSSMAN, R. C. 1896 The meteorology of Edinburgh. *Transactions of the Royal Society of Edinburgh*, **39**, 63–207.

NKEDERIM, L. C. 1988. An assessment of the relationship between functional groups of weather elements and atmospheric pollution in Calgary Canada, *Atmospheric Environment*, **25a**, 2287–2296.

PARKER, D. E., LEGG, T. P. & FOLLAND, C. K. 1992. A new central England temperature series. *International Journal of Climatology*, **12**, 317–342.

POLLACK, J. B., TOON, O. B., SAGAN, C., SUMMERS, A., BALDWIN, B. & VAN CAMP, W. 1976. Volcanic eruptions and climatic change: a theoretical assessment. *Journal of Geophysical Research*, **816**, 1071–1083.

PREINING, O. 1991. Aerosols and Climate – An overview. *Atmospheric Environment*, **25A11**, 2443–2444.

—— 1992. Global warming: greenhouse gases versus aerosols. *Science of the Total Environment*, **126**, 199–204.

RAMPINO, M. R. 1988. Volcanic winters. *Annual Review of Earth and Planetary Science*, **16**, 73–99.

—— STOTHERS, R. B. & SELF, S. 1985. Climatic effects of volcanic eruptions. *Nature*, **313**, 272.

ROBOCK, A. & MAO, J. 1992. Winter warming from large volcanic eruptions. *Geophysical Research Letters*, **24**, 2405–2408.

SEAR, C. B., KELLEY, P. M., JONES, P. D. & GOODESS, C. M. 1987. Global surface temperature responses to major volcanic eruptions. *Nature*, **330**, 365–367.

SIGURDSSON, H. 1982. Volcanic pollution and climate. *Eos*, **6332**, 601–603.

—— 1990. Evidence of volcanic loading of the atmosphere and climatic response. *Palaeogeography, Palaeoclimatology, Palaeoecology*, **89**, 277–289

STOTHERS, R. B. 1996. The Great Dry Fog of 1783. *Climatic Change*, **32**, 79–89.

—— & RAMPINO, M. R. 1983. Volcanic eruptions in the Mediterranean before A.D. 630 from written and archaeologcal sources. *Journal of Geophysical Research*, **88B8**, 6357–6371.

THORARINSSON, S. 1969. The Lakagigar eruption of 1783. *Bulletin Volcanologique*, **333**, 910–921.

THORDARSON, TH, & SELF, S. 1993. The Laki Skaftar fires and Grimsvotn eruptions in 1783–85. *Bulletin of Volcanology*, **55**, 233–263.

TRIPOLI, G. J. & THOMPSON, S. L. 1988. A three-dimensional numerical simulation of the atmospheric injection of aerosols by a hypothetical basaltic fissure eruption. *Global Catastrophes and Earth History, Abstract Volume LPI and NAS conference, Snowbird, Utah, 20–23 Oct 1988*, 200–201.

WAKAMATSU, S., UNO, I. & SUSUKI, M. 1990. A field study of photochemical fog formation under stagnant meteorological conditions. *Atmospheric Environment*, **24A5**, 1037–1050.

WHITE, G. 1789. *The Natural History of Selbourne* (reprinted 1977). Penguin Books, London.

WOOD, A. 1984. The amazing and portentous summer of 1783. *Eos*, **65**, 410–411.

—— 1992. Climatic effects of the 1783 Laki eruption. *In:* HARINGTON, C. R. (ed.) *The Year Without a Summer*, Canadian Museum of Nature, 58–77.

WOODFORDE, J. 1984. *The Diary of a Country Parson 1758–802*. Oxford University Press.

Environmental and social responses in Europe to the 1783 eruption of the Laki fissure volcano in Iceland: a consideration of contemporary documentary evidence

MARK BRAYSHAY[1] & JOHN GRATTAN[2]

[1] *The University of Plymouth, Department of Geographical Sciences, Drake Circus, Plymouth, Devon PL4 8AA, UK*
[2] *The University of Wales, Institute of Geography and Earth Sciences, Llandinam Building, Aberystwyth, Dyfed SY23 3DB, UK*

Abstract: A detailed examination of contemporary documentary evidence, including letters, diaries, historical accounts and newspaper reports, reveals the dramatic effect on the weather across the whole of western Europe of the eruption of the Laki volcanic fissure in Iceland in 1783. Extreme heat, dry sulphurous fogs, chemical pollution, and tremendous storms of thunder, lightning and hail were reported from northern Scotland to Sicily. Vegetation was defoliated, crops were destroyed, livestock were killed and property was damaged. There were also direct and indirect human casualties. The unusual conditions engendered considerable fear as well as an appeal to science for a rational explanation. Volcanic eruptions and earthquakes in southern Italy and Iceland were blamed as the cause.

Recent advances in both field and laboratory techniques have made it possible to detect with considerable accuracy the spatial dispersion of volcanic ejecta produced during past eruptive episodes (Dugmore 1989; Dugmore & Newton 1992). Moreover, by analysing the chemical signature of surviving tephra sherds, it is now feasible to establish a link between particular deposits and known volcanic eruptions (Bennett *et al.* 1992; Pilcher & Hall 1992; Pilcher *et al.* 1995). While there are dangers in attributing past changes in settlement and culture over wide areas entirely to the impact of volcanic activity, populations and landscapes located close to eruptions are known to have been affected in a catastrophic manner (Jashemski 1979; Renfrew 1979; Albore-Livadie, 1980), and recent work has shown how evidence of such extreme impacts may be readily detected in the archaeological record (Torrence *et al.* 1990; Grattan & Brayshay 2000). However, in locations distant from a volcano an eruption can rarely, if ever, be observed to have caused wholesale destruction and the precise character of such distal impacts is difficult to gauge (Grattan 1994; Grattan & Gilbertson 1994). Archaeological evidence is often not sufficiently sensitive to identify subtle or short-term impacts and palaeoenvironmental investigations may be contradictory (Blackford *et al.* 1992; Hall *et al.* 1994). Thus, although the presence of tephra deposits may confirm that volcanic pollutants reached an area, unless these were on a scale large enough to precipitate severe levels of depopulation and abandonment, the precise character of the environmental and social responses cannot be easily reconstructed. Palaeoenvironmental and archaeological studies are therefore unable to model fully the impact of volcanic eruptions on distant societies and ecosystems. However, a study of available archival material can offer an opportunity to reconstruct in detail the environmental and social effects of volcanic ejecta and the atmospheric disturbances associated with the particulate and gaseous pollutants of historically famous eruptions, and provide a theoretical framework on which to model the impact of temporally remote events.

This paper seeks to explore the use of historical documentary evidence in the reconstruction of responses across western Europe to the eruption of the Laki fissure volcano in 1783. The potential for building a significant corpus of corroborative descriptions in contemporary letters, diaries, historical accounts and newspaper reports of the climatic and social responses triggered by the presence of volcanic material in the atmosphere will be demonstrated.

Eruption dynamics and proximal impacts

The 1783 eruption of the Laki fissure in Iceland is regarded as one of the most important volcanic episodes of the modern era (Thordarson & Self 1993). Commencing on 8 June 1783, and lasting until the following February, during its eight months of activity Laki produced 9.9×10^{13} g of acid gases (mostly H_2SO_4). Indeed, about 60% of the acid gases were discharged from the fissure in five concentrated episodes between 8 June and 8 July, 1783. Thus, while the 1783 eruption of Laki has been assigned a volcanic explosivity index (VEI) of only 4, it nonetheless yielded a quantity of gases and aerosols on a scale rarely approached by other Holocene northern hemisphere volcanic eruptions (Newhall & Self 1982; Devine et al. 1984; Sigurdsson et al. 1985; Palais & Sigurdsson 1989; Thordarson & Self 1993). Moreover, lacking the explosive energy of other eruptions which engendered sufficient power to force ejecta directly into the stratosphere, an estimated two-thirds of the gases emitted by Laki in 1783 are thought to have been confined to the troposphere (Thordarson & Self 1993; Fiacco et al. 1994). The evidence assembled by Kington (1988), who has reconstructed a detailed picture of the weather in Europe throughout the 1780s, indicates that there was a high pressure cell located over northern Europe and a low pressure cell located over Iceland in late-June and July 1783. It may be argued that the volcanic gases and tephra ejected from Laki were drawn to a high altitude within the troposphere and then rapidly conveyed by high velocity winds towards the high pressure cell stationed over Europe, where they descended quickly towards the land surface (Stothers 1996). Moreover, Kington's sequence of daily weather maps for 1783 shows that the centre of the high pressure cell shifted its position during the first few weeks of July 1783: at first it moved northwards to Scandinavia, but then drifted southwards to the Alps and northern Italy. This implies that volcanic aerosols and tephra from Laki were disbursed across the entire continent. A fuller consideration of the climatological conditions which prevailed in Europe during the summer of 1783 is provided elsewhere in this volume (Grattan & Sadler 1999).

In Iceland itself, research has indicated considerable damage to the physical environment and high levels of mortality in the wake of the eruption of Laki. Pasture was destroyed and large numbers of sheep, cattle and horses perished, and an estimated 24% of the human population subsequently died (Thórarinsson 1979;

Hálfdanarson 1984; Pétursson *et al.* 1984; Ogilvie 1986). Other investigations have indicated that crops were damaged in Scandinavia and an ash fall occurred in northern Scotland (Giekie 1893; Lamb 1970; Thórarinsson 1981), yet very little attention has so far been paid to its impact in other parts of Europe (Grattan & Brayshay 1995).

Documentary evidence of environmental and social responses to the Laki eruption

An extensive search was undertaken of personal letters, diaries and journals, topographical accounts and contemporary newspaper reports which refer to highly disturbed weather events occurring in the summer of 1783. A large body of descriptive information regarding strange weather phenomena across wide areas of Europe was assembled (Fig. 1). It should be noted that the majority of the sources used in this research were drawn from Great Britain and we are thus reporting environmental impacts experienced in Europe viewed from a British perspective. A further development of this work has since been to consult comparable sources in France, Spain, Italy and Germany with the object of extending and amplifying the British evidence (Grattan *et al.* 1998).

Surviving diaries, journals and letters were found frequently to provide not only eyewitness accounts of environmental phenomena, but also comments on popular reactions to them. The task of assembling descriptive accounts of this kind is, of course, formidable and it is likely that many more remain to be consulted (Matthews 1950; Gard 1989). Moreover, an incidental reference to a violent storm of thunder and lightning occurring in July or August 1783, discovered in a personal diary, does not necessarily, by itself, indicate a direct link with the Laki eruption. However, when such reports are examined within the context of other similar evidence, and corroborated over a wide area, a pattern of extraordinarily disturbed atmospheric conditions emerges.

An additional source consulted in this research was the large number of reports on disturbed weather, and reactions to it, contained in Britain's newspapers (a list of the key titles searched for evidence of this kind is given in Table 1). By the 1780s, both the London and provincial newspapers were enjoying a significant growth in their readership and geographical circulation (Cranfield 1962, 1978). Although the local press continued to rely heavily upon news published in the London papers, increasing space was devoted to reports of events occurring in their own and other areas (Wiles 1965; Murphy 1972; Feather 1985). However, it was not normal in the eighteenth century to include reports on the weather and atmospheric conditions. Thus, it is particularly significant that, beginning in June 1783, extremely unusual weather events were increasingly reported in newspapers published throughout Great Britain (Grattan & Brayshay 1995). The unprecedented character of the climatic disturbances then being experienced appears to have prompted an interest in whether other parts of Europe were similarly affected. Indeed, amongst almost 200 documentary references drawn from letters, diaries, journals, eye-witness reports and newspapers which record unusual weather phenomena, at least 30 focus exclusively on the conditions experienced elsewhere in Europe. Others draw comparisons with Europe or make incidental reference to similar conditions prevailing on the continent.

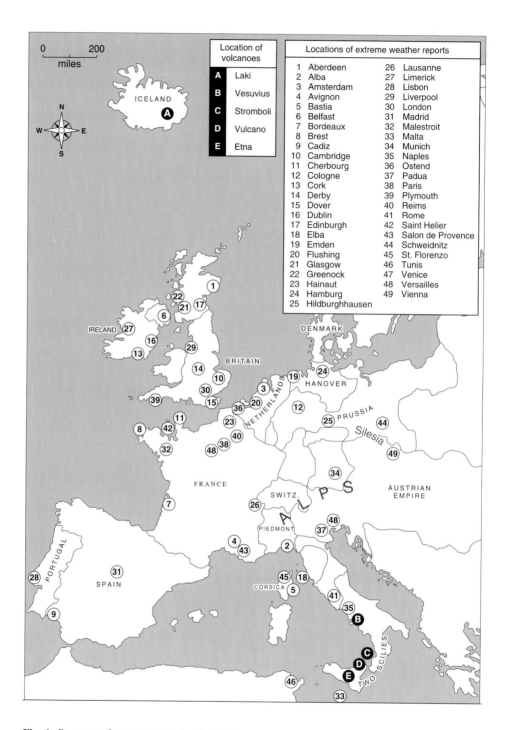

Fig. 1. Reports of extreme weather in 1783.

Table 1. *Newspapers consulted in the research*[a]

Aberdeen Journal [12]	7 July–1 September
Bristol Journal [9]	12 July–23 August
Bury Post and Universal [19]	19 June–3 September
Caledonian Mercury [11]	23 June–25 August
Cambridge Chronicle and Journal [6]	5 July–23 August
Edinburgh Advertiser [21]	4 July–19 August
Exeter Flying Post [9]	3 July–28 August
General Evening Post [12]	5 July–31 July
Ipswich Journal [12]	12 July–30 August
London Gazette [2]	17 July–30 August
London Packet [14]	25 June–29 August
Morning Chronicle and London Advertiser [6]	4 July–20 August
Morning Herald and Daily Advertiser [10]	11 July–26 August
Parker's Advertiser and Morning Intelligence [1]	11 July
Public Advertiser [4]	7 July–17 July
Norfolk Chronicle [4]	12 July–23 August
Sherborne Mercury [9]	14 July–21 July
York Courant [17]	10 June–26 August
Whitehall Evening Post [14]	21 July–16 August

Information was gathered from all of the newspapers listed and relevant articles were found for 1783 during the time periods indicated. The most relevant reports commenced on 10 June and continued until the first week of September, 1783. The number of reports appearing in each title is given in brackets. Bold typeface indicates the titles included significant reports drawn from other parts of Europe.
[a] Note: The majority of these titles were consulted in the National Newspaper Library (British Library), Colindale, London; some were also found to be available in various county local studies libraries.

Hot, malodorous fogs linked with volcanic pollution in Europe in the summer of 1783

The summer of 1783 in Europe is noted for the persistence of a hot, smoky, sulphurous fog which obscured the sun and caused a considerable reduction in visibility (Fig. 2). The English naturalist, Gilbert White, commented:

> the peculiar haze or smoky fog, that prevailed for many weeks in this island *and in every part of Europe*, and even beyond its limits, was a most extraordinary appearance, unlike anything known within the memory of man (White 1789 – our italics).

The poet and essayist, William Cowper, was also struck by the density and persistence of the fog and remarked that:

> ... so long in a country not subject to fogs, we have been cover'd with one of the thickest I remember. We never see the sun but shorn of his beams, the trees are scarce discernible at a mile's distance, he sets with the face of a hot salamander and rises with the same complexion (Cowper, 29 June 1783, in King & Ryskamp 1981).

Other diarists, including the Revd James Woodforde of Weston in Norfolk, noted the unusual heat and redness of the sun (25 June 1783, in Beresford 1924–31). William Dunn, a Worcestershire schoolmaster, noted on 10 July 1783 that 'the air continues in a putrid state; the sun is as red as blood' (Bodleian MS, Don. c. 76). Indeed, the reddish, malodorous, dry fogs which prevailed across Europe for many weeks in 1783 were not dissipated, even by the intense heat (Orlandini 1853). In his detailed annual accounts of the weather, William Gilpin noted that in 1783, 'during almost all the summer months, the sky was overspread with a dark dry fog'. Gilpin

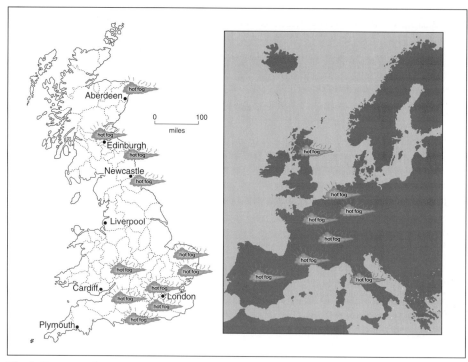

Fig. 2. Reports of 'Hot Fogs' in Britain and Europe, 1783.

also reported that the:

> gloomy atmosphere was soon found to be general all over England. Advices from the continent: France, Spain, Italy and other parts, showed it had equally overspread all the countries of Europe and by degrees it was found to be universal over the face of the globe (Bodleian MS, Eng. Misc. d. 564).

Camuffo & Enzi (1995) in a study of the impact of clouds of volcanic aerosols in Italy during the past 700 years, indicate that the fog experienced in the Veneto and Lombardy in 1783 was due to the Laki eruption and cite the report of a contemporary Italian chronicler, Gennari, who noted that for several weeks the fog had persisted from dawn until sunset and:

> ... the moon appeared ruddy and ... the sun could be looked at without being blinded. The fog was hot, dry and dense, and this phenomenon was observed not only by us, but also elsewhere in Italy, Germany and France (Gennari 1783, in Camuffo & Enzi 1995).

Benjamin Franklin, who was at that time serving as the American ambassador in Paris, noted that:

> during several of the summer months of the year 1783, when the effect of the sun's rays to heat the earth in these northern regions should have been greatest, there existed a constant fog over all Europe ... this fog was of a permanent nature; it was dry, and the rays of the sun seemed to have little effect towards dissipating it (Franklin 1784).

Reports from Paris mentioned 'a very unusual degree of heat in the air' together with a 'fog, haziness and redness of the sun, very remarkable' (*Edinburgh Advertiser*, 15 July 1783). The *Bristol Journal* (19 July 1783) noted that 'those who are lately come from Rome say it is as thick and hot in Italy, and that even the top of the Alps is covered with it, and travellers and letters from Spain affirm the same'. Information from Naples, dated 15 July, noted the continuing fogs which were accompanied by:

> ... so alarming an increase of obscurity that our bargemen do not dare venture on the waters without compass (*Morning Herald and Daily Advertiser*, 19 August 1783).

A report from Salon de Provence in the south of France, dated 11 July, spoke of twenty days of continuous fog:

> The sun, although hot, does not dissipate it, neither day nor night ... the countryside appears whitish grey ... and the fog sometimes emits a strong odour and is so dry, it does not tarnish a looking glass and instead of liquifying salts, it dries them (*Aberdeen Journal*, 18 August 1783).

These sultry, polluted conditions appear to have increased the scourge of infectious disease, particularly in the crowded, unhealthy towns of southern Europe, where epidemics were said to have been worse than usual in the summer of 1783 (Camuffo & Enzi 1995). Indeed, in southern Italy, the polluting effect of the persistent hot dry fog may have been further reinforced by emissions from Stromboli, Vulcano and Vesuvius which were all active in 1783 (Simkin *et al.* 1981). The British press reported considerable seismic activity in Calabria during June and July in the wake of the eruptions of the Italian volcanoes (*York Courant*, 1 , 22 & 29 July, 5, 12 & 19 August, 1783; *Whitehall Evening Post*, 7 & 12 August, 1783). Contemporary Italian observers were persuaded that the 'dust agitated by the violent shocks of the earth' was a factor contributing to the persistent dry mist which blanketed southern Italy and Sicily (Toaldo 1784).

Reports from Hildburghhausen (Fig. 1) described a thick sulphurous vapour, but attributed this to an eruption of Mount Gleichberg (*Morning Herald and Daily Advertiser, Whitehall Evening Post*, 12 August 1783). However, reports from Emden acknowledge that:

> the thick dry mist which has continued so long, seems spread over the whole of Europe ... during the day it veils the sun and in the evening there is a tainted odour (*Morning Herald and Daily Advertiser*, 5 August 1783).

The evidence suggests that damage to vegetation associated with the hot fogs was considerable. Accounts from northern Germany and the Netherlands report not only the 'infectious smell' of the fog, but also that 'all the trees on the borders of the Ems have been stripped of their leaves in one night' (*Ipswich Journal*, 9 August 1783).

Further damage to plants appears to have occurred as a result of brief, severe episodes of cold. The hot fogs caused considerable evaporation from the land surface and the moisture taken up in this way formed dense masses of clouds. During this process the temperature could suddenly drop and, although it rose again almost immediately, the effect on crops could be devastating. In England it was reported that:

> throughout most of the eastern counties there was a most severe frost between the 23rd

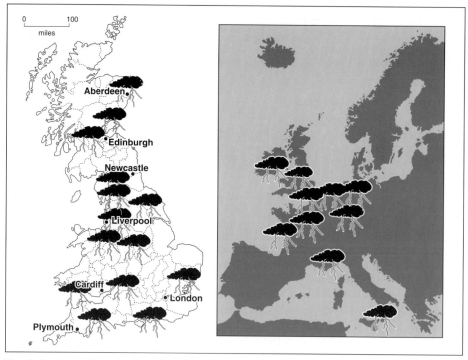

Fig. 3. Reports of tremendous thunderstorms in Britain and Europe, 1783.

and 24th June. It turned most of the barley and oats yellow, to their very great damage; the walnut trees lost their leaves and the larch and firs in plantations suffered severely (*Sherborne Mercury*, 14 July 1783).

The synergistic relationship of frosts, fogs, atmospheric pollution and crop damage in 1783 has been explored and described by Grattan & Charman (1994). Similar experiences appear to have been shared elsewhere in Europe (Orlandini 1853).

Violent storms of thunder, lightning and hail

The high levels of evaporation from the surface induced by the hot fogs charged the atmosphere with water vapour. Moreover, the presence in the air of large quantities of hygroscopic particles of volcanic origin provided highly efficient condensation nuclei. The result was a spate of ferocious storms of thunder and lightning which swept Europe between June and September 1783. The Revd John Russell Greenhill of Fringford in Oxfordshire recorded several episodes of 'violent thunder, lightning and rain' in his diary during July 1783 (Bodleian MS, Eng. Misc. e. 200). Writing his *History of Devonshire* in 1793, the Revd Richard Polwhele looked back on 1783 and reported that:

> Perhaps a more troubled atmosphere than we felt in 1783 was never experienced in any former year. The whole summer was tempestuous. On a high down near Colyton, I had a very narrow escape from what appeared like a ball of fire, glittering as it passed in a line parallel with the head of my horse. At the moment in which the electric fluid was

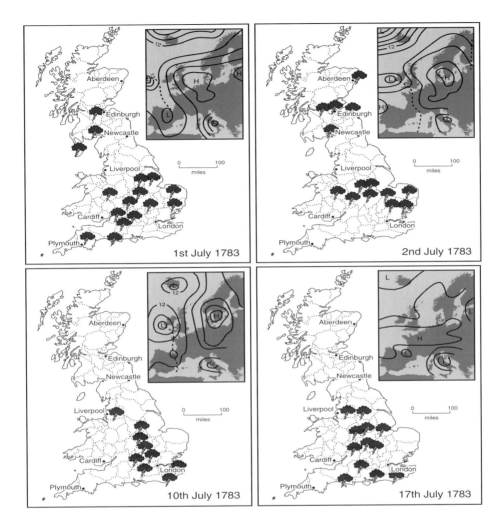

Fig. 4. The passage of tremendous thunderstorms across Britain in July 1783. Synoptic maps adapted from Kington (1988).

discharged from the cloud, the whole country around me seemed to receive the shock as of an earthquake. In the same year fell a storm at Umberleigh and the neighbourhood, that set on fire several houses and barns, destroyed all the corn where it extended, ploughed up the road in many places ten feet deep, and rooted up large trees (Polwhele 1793).

The experiences of this Devonshire clergyman were similar to those of people throughout Europe (Fig. 3). The *London Packet* (27 June 1783) reported that the 'greater part of the [Corsican] town of St Florenzo has been destroyed by fire which was occasioned by a storm of thunder and lightning'. On a night in the last week of June, across the whole of western Ireland, but especially in the city and county of Cork, there was 'a most tremendous storm of thunder and lightning' (*Caledonian*

Mercury, 2 July 1783). Within days there were reports of storm damage to property on the island of Elba (*Caledonian Mercury*, 9 July 1783). In fact this disturbed weather continued for many weeks. Reports from Cherbourg spoke of the 'most violent storms of thunder and lightning ever known in those parts', while at Schweidnitz or Swidnica in Poland there had been 'so dreadful a storm that there was no distinguishing it from an earthquake' (*York Courant*, 22 July 1783). Letters from Amalfi in Italy reported frequent and destructive thunderstorms in July and August (*Morning Herald and Daily Advertiser*, 19 August 1783). The *Cambridge Chronicle and Journal* (23 August 1783) described violent storms at Malestroit and Hainaut in France (Hainaut is now in Belgium). The same report mentioned a violent hailstorm at Reims where 'many of the hail stones were as large as a hen's egg'. On 5 August, a storm of thunder, hail and lightning in Amsterdam was reported as one of the worst ever experienced (*York Courant*, 19 August 1783). Although the storms were ferocious, it was frequently noted that they seldom cleared the air for long of the sulphurous fogs, and rarely reduced the high temperatures:

> the storm of thunder and lightning on Thursday evening last ... came with the wind over the channel from France ... [and] was violently alarming; the flashes of lightning were remarkably sulphureous, and the peals of thunder loud and awful ... As the storm came on in most places the thermometer kept rising. Where the rain fell, the thunder was most violent, where there was no rain the sulphureous stench remained in the air the greater part of the next day, when the heat was more intense than before (*Sherborne Mercury*, 21 July 1783).

By assembling a large number of documentary references to unusual climatic phenomena for Britain, it has been possible to reconstruct detailed maps of particular violent storms of thunder and lightning. Thus the precise locations where the sequence of storms occurred on 1, 2, 10 and 17 July 1783 have been identified (Fig. 4). The patterns thereby revealed show a convincing degree of agreement with the weather charts constructed by Kington (1988). As already noted, his work shows clearly that, during July, the centre of the large high pressure system over Europe drifted a little: first in a northwards direction towards Scandinavia, but then returned southwards to a position, by the end of the month, over the Alps and northern Italy. Charged with acid aerosols and other volcanic volatiles ejected by Laki, these movements of the high pressure cell would have ensured that pollution was delivered to virtually the entire continent (Stothers 1996). Given these important atmospheric controls which ensured the effective horizontal spread of Laki pollution, the considerable scope to extend further the investigation of its impact across the whole continent of Europe is now beginning to be realized (Grattan *et al.* 1998, 1999).

Besides damage to buildings, crops were destroyed and both livestock and people were injured or killed. At Belper in Derbyshire on 7 August there was a typically violent hailstorm:

> The stones were extremely large, some of them being measured were found to be from four to six inches in circumference, and of various forms such as round, oblong, square &c. &c. This hail did damage in the direction it fell, to the standing corn, and some windows ... People selling pots, in Belpar market, had them broke in the same manner as if stones had been thrown upon them (*Exeter Flying Post*, 14 August 1783)

The *London Gazette* (21 August 1783) reported storms in the Avignon area of southern France where 'storms multiplied of late and have done great damage in country places'. On the 2 July there had been a storm at St Esprit 'which destroyed the harvest ... the hailstones were the size of hens' eggs and their irregular form cut the vines and trees'. An unfortunate young girl was struck on the head by a hailstone and 'she found that her cap was on fire'. The same report mentioned balls of fire and huge hailstones during a 'terrible storm in Bouches du Rhône'.

Ball lightning of the kind that narrowly missed the Revd Polwhele's horse in Devon was reported not only elsewhere in Britain, but also in many places in Europe. The damage could be severe. Houses and barns were struck and set on fire, crops and livestock were destroyed. In county Cork 'balls of fire from the lightning were seen to fall in many fields, tearing the ground as they entered' (*Caledonian Mercury*, 13 August 1783). Even the palace of Louis XVI of France sustained storm damage; letters received by the French mails reported that they had:

> ... just received melancholy accounts from Versailles of great damage being done by thunder and lightning in that town ... part of the King's palace was shattered to pieces, and some of the horses struck dead ... houses were unroofed and many lives lost (*Edinburgh Advertiser*, 29 July 1783).

Social responses to the extreme weather events

In a phrase which encapsulates the mood across Europe in response to the violent episodes of weather that were sweeping the continent, Gilbert White characterized the summer of 1783 as an 'amazing and portentous one, and full of horrible phenomena ... that affrighted many counties of this kingdom' (White 1789). In rather more sanguine language, Richard Polwhele spoke of a 'tempestuous' summer when the atmosphere was 'more troubled than in any former year' (Polwhele 1793). There is, however, little doubt that the persistent fogs, the stench of sulphur in the air and, above all, the tremendous storms, caused considerable alarm. The following description of responses to the episode of thunder and lightning which occurred on 1 July in Gloucester, sums up popular reactions:

> The rain fell in heavy drops as large as walnuts ... the lightning so strong, and thunder so tremendous, that the women, shrieking and crying, were running to hide themselves, the common fellows fell down on their knees to prayers, and the whole town was in the utmost fright and consternation (*Exeter Flying Post*, 10 July 1783).

Coupled with fear, was a sense of wonder at the sheer magnificence and ferocity of the storms. Reporting the arrival of the storm of 2 July in Leicester, the *Bristol Journal* (19 July 1783) gave the following account of the scene of 'inconceivable horror':

> It first began about nine in the evening at a distance, westerly, the sky clear, the atmosphere uncommonly hot and sultry; by degrees the clouds and lightning approached, exhibiting a wonderful spectacle of dreadful magnificence; before eleven o'clock the whole firmament appeared all on fire; the thunder, which until now had been heard at a distance, became loud and terrible, and the rain descended in torrents, mixed with smoke, sulphur and liquid fire.

The extraordinary character of the weather was thought to be unprecedented. In Paris, relentlessly high temperatures were blamed for a spate of 'canine madness' and gentlemen 'walking in the Tuileries' park were apparently attacked by dogs driven mad in the extreme heat (*London Packet*, 27 June 1783). Such conditions caused consternation in France among:

> the superstitious part of the people, who had been wrought upon by their priests to believe that the end of the world was at hand (*Edinburgh Advertiser*, 18 July 1783).

Appeals were made to scientists to explain the unusual phenomena. In response, a group of physicists of the French Academy of Sciences, intent on making some observations of the atmosphere, proceeded to the Paris Observatory where they:

> had a sort of kite flown to a great height, after which it was drawn in covered with innumerable small black insects which, upon examination, appeared to contain moisture prejudicial to plants (*Bury Post and Journal*; *London Gazette*, 31 July 1783).

Reassurances were offered by the French astronomer, De La Lande, who published a short paper to 'quiet the minds of the people' (*Edinburgh Advertiser*, 18 July 1783). He acknowledged that the 'multitude draw strange conclusions when they see the sun of a blood colour, shed a melancholy light, and cause a musty sultry heat'. But, he argued, that this was nothing more than:

> ... a very natural effect from a hot sun after a very heavy succession of rain. The first impression of heat has necessarily and suddenly rarefied a superabundance of watery particles with which the earth was deeply impregnated, and given them, as they rose, a dimness and rarefaction not unusual to common fogs. This effect, which seems to me very natural, is not so very new; it is at most not above nineteen years since there was a like example, which period brings the moon in the same position on the same days, and which appears to have some influence on the seasons ... Among the meteorologic observations of the Academy for the month of July 1864, I find ... a great resemblance to the latter end of our June, so that it is not an unheard of or forgotten thing. In 1764, they had afterwards storms and hail, and nothing worse need be feared in 1783 (reported in: *Bristol Journal*, 19 July 1783).

Despite the inquiries by the scientists, few people in 1783, with the exception of Benjamin Franklin (1784), appear to have attributed Europe's disturbed weather to the volcanic activity in Iceland. Horace Walpole linked the unusual weather in England to the volcanic activity reported from Italy:

> I am quite persuaded ... that the dreadful eruptions of fire on the coasts of Italy and Sicily should have occasioned some alteration that has extended faintly hither, and contributed to the heats and mists that have been so extraordinary (Walpole, 15 July 1783, in Cunningham 1938).

Speculating on what had caused 'these heavy fogs' that had 'overspread the earth', William Gilpin noted that:

> Many were of the opinion that they were the effluvia of those dreadful earthquakes which had desolated Calabria ... But how that cause was adapted to produce such an effect, was another question ... Never was remembered a year more remarkable for ... thunderstorms; which indeed made it still more probable that these vapours originated in the sulphurious steams of earthquakes and volcanoes (Bodleian MS, Eng. Misc. d. 564).

While an understanding of the origins of the tempestuous conditions may have eluded those who witnessed them, the documentary evidence leaves no doubt that, for Europe, the summer of 1783 was indeed 'amazing and portentous':

> All accounts from every part of Europe agree, this has been the driest and hottest summer that has been experienced for many years. Storms in France, earthquakes in Italy, and the plague in Turkey, are three national visitations which have not been felt in such extremes for a long period of time (*Exeter Flying Post*, 28 August 1783).

Conclusion

Interdisciplinary studies in which the skills of a historian are combined with those of a physical scientist can make a significant contribution to an understanding of the environmental and social impact of a volcanic eruption occurring some two centuries ago, which did not cause wholesale destruction beyond its immediate vicinity, but nonetheless polluted the atmosphere with acid gases, aerosols and tephra over a wide tract of Europe for a period of several months. This paper has shown how a detailed search of historical documentary evidence can be used in reconstructing the pattern and character of responses across western Europe to the eruption in Iceland of the Laki fissure volcano in 1783. Contemporary letters, diaries, historical accounts and newspaper reports have shed much new light on both the incidence and the consequences of the extreme heat, the air pollution, the foul-smelling and persistent fogs, crop damage, social unrest and the spectacular storms triggered by the presence of volcanic material in the atmosphere. No claim is made that all the available sources of evidence have yet been recovered. Moreover, the findings reported here rely heavily on documentary material surviving in Britain and there remains considerable scope for extending this research by examining comparable European sources, thereby corroborating more fully the phenomena described in this paper.

The authors would like to thank Tim Absalom and Andrew Hoggarth for preparing the figures.

References

ALBORE-LIVADIE, C. 1980. Palma Campania (Napoli) – Resti di abitato dell'eta del bronzo antico. *Atti della Academia nazionale dei Lincei*, **XXIV**, 59–101
BENNETT, K. D., BOREHAM, S., SHARP, M. J. & SWITSUR, V. R. 1992. Holocene history of environment, vegetation and human settlement of Catta Ness, Lunnasting, Shetland. *Journal of Ecology*, **80**, 241–73.
BERESFORD, J. 1924–31. *The diary of a country parson, 1758–1802*. 5 Vols. OUP, Oxford.
BLACKFORD, J. J., EDWARDS, K. J., DUGMORE, A. J., COOK, G. T. & BUCKLAND, P. 1992. Hekla-4: Icelandic volcanic ash and the mid-holocene Scots Pine decline in northern Scotland. *The Holocene*, **2**(3), 260–265.
CAMUFFO, D. & ENZI, S. 1995. Impact of the clouds of volcanic aerosols in Italy druing the last seven centuries. *Natural Hazards*, **11**, 135–161.
CRANFIELD, G. A. 1962. *The development of the provincial newspaper, 1700–1760*. Clarendon Press, Oxford.
—— 1978. *The press and society: from Caxton to Northcliffe*. Longman, London.
CUNNINGHAM, P. 1938. *The letters of Horace Walpole*. (Vol. VIII). Richard Bentley, London.
DEVINE, J. D., SIGURDSSON, H., DAVIS, A. N. & SELF, S. 1984. Estimates of sulphur and chorine yield to the atmosphere from volcanic eruptions and potential climatic effects. *Journal of Geophysical Research*, **89**, 6309–6325.
DUGMORE, A. J. 1989. Icelandic volcanic ash in Scotland. *Scottish Geographical Magazine*.

105(3), 168–172.

—— & NEWTON, A. J. 1992. Thin tephra layers in Peat revealed by X-radiography. *Journal of Archaeological Science*, **19**, 163–170.

DUNN, W. *Diary, with some accounts, of William Dunn, schoolmaster of Belbroughton, Worcs., 1767–95, including daily reports on the weather*. Bodleian MS, Don. c. 76.

FEATHER, J. 1985. *The provincial book trade in eighteenth-century England*. Cambridge University Press, Cambridge.

FIACCO, R. J., THORDARSON, TH., GERMANI, M. S., SELF, S., PALAIS, J., WHITLOW, S. & GROUTES, P. 1994. Atmospheric aerosol loading and transport due to the 1783–84 Laki fissure eruption in Iceland. Interpreted from Ash particles and acidity in the Gisp2 ice-core. *Quaternary Reviews*, **42**(3), 231–244.

FRANKLIN, B. 1784. Meteorological imaginations and conjectures. *Memoirs of the Literary and Philosophical Society of Manchester*, **2**, 373–377.

GARD, R. 1989. *The observant traveller: diaries of travel in England, Wales and Scotland in the county record offices of England and Wales*. HMSO, London.

GIEKIE, SIR A. 1893. *Text book of geology*. Macmillan, London.

GILPIN, W. *An historical account of the weather during twenty years from 1763–1785*. Bodleian MS, Eng. Misc. d. 564.

GRATTAN, J. P. 1994. Land abandonment and Icelandic volcanic eruptions in the late 2nd millennium BC. *In:* STÖTTER, J. & WILHELM, F. (eds), *Environmental Change In Iceland*, Münchener Geographische Abhandlungen, B12, 111–132.

—— & BRAYSHAY, M. 1995. An amazing and portentous summer: environmental and social responses in Britain to the 1783 eruption of an Iceland volcano. *Geographical Journal*, **161**(2), 125–134.

—— & —— 2000. Modelling the impact of the Vesuvius–Avellino eruption upon the Bronze Age settlement of Palma Campania. *PACT*, in press.

——, —— & SADLER, J. 1998. Modelling the distal impacts of past volcanic gas emissions: Evidence of Europe-wide environmental impacts from gases emitted during the eruption of Italian and Icelandic volcanoes in 1783. *Quaternaire*, **9**, 25–35.

—— & CHARMAN, D. J. 1994. The palaeoenvironmental implications of documentary evidence for impacts of volcanic volatiles in western Europe, 1783. *The Holocene*, **4**(1), 101–106.

—— & GILBERTSON. D. D. 1994. Acid-loading from Icelandic tephra falling on acidified ecosystems as a key to understanding archaeological and environmental stress in northern and western Britain. *The Journal of Archaeological Science*, **21**(6), 851–859.

—— & SADLER, J. 1999. Regional warming of the lower atmosphere in the wake of volcanic eruptions: the role of the Laki fissure eruption in the hot summer of 1783. *This volume*.

——, SCHÜTTENHELM, R. T. E. & BRAYSHAY, M. 1999. Severe social and environmental responses to volcanic pollution of the lower atmosphere: an alternative forcing mechanism for archaeologists and historians. *In:* MCCOY, F. W. & HEIKEN, G. (eds) *Volcanic Hazards and Disasters in Human Antiquity*. Geological Society of America, Special Publications, in press.

GREENHILL, J. R. *Original diary of Rev. John Russell Greenhill DCL, Rector of Marsh Gibbons, Bucks. & of Fringford, Oxon. from 1 January 1780 to 31 August 1787 with the daily state of the atmosphere & weather during that period from personal observation*. Bodleian MS, Eng. Misc. e. 200.

HÁLFDANARSON, G. 1984. Mannfall I Móouharoindum. *In:* GUNNLAUGASSON, G., GUOBERGSSON, G., PÓRARINSSON, S., RAFNSSON, S & EINARSSON, P. (eds) *Skaftáreldar*. Mál Og Menning, Rekyavik, 139–162.

HALL, V. A., PILCHER, J. R. & MCCORMAC, F. G. 1994. Icelandic volcanic ash and the mid-Holocene Scots pine (PINUS SYLVESTRIS) decline in the north of Ireland: no correlation. *The Holocene*, **4**(1), 79–83.

JASHEMSKI, W. F. 1979. Pompeii and Mount Vesuvius, A.D. 79. *In:* Sheets, P. D. and Grayson, D. K. (eds) *Volcanic activity and human ecology*. Academic Press, New York and London, 587–621.

KING, J. & RYSKAMP, C. 1981. *The letters and prose writings of William Cowper*. (Vol II). Clarendon Press, Oxford.

KINGTON, J. A. 1988. *The weather patterns for the 1780s over Europe.* Cambridge University Press, Cambridge.

LAMB, H. H. 1970. Volcanic dust in the atmosphere, with a chronology and assessement of its meteorological significance. *Philosophical Transactions of the Royal Society of London, Series A* **266** (1170), 425–533.

MATTHEWS, W. 1950. *British diaries, 1442–1942.* University of California Press, Berkeley.

MURPHY, M. J. 1972. Newspapers and opinion in Cambridge, 1780–1850. *Transactions of the Cambridgeshire Bibliographical Society*, **VI**, 35–55.

NEWHALL, G. C. & SELF, S. 1982. The Volcanic Explosivity Index (VEI): an estimate of explosive magnitude for all volcanism. *Journal of Geophysical Research*, **87**(C2), 31–38.

OGILVIE, A. E. J. 1986. The climate of Iceland 1701–1784. *Jökull*, **36**, 57–73.

ORLANDINI, O. 1853. *Trattato completo di Meteorologia Agricola.* Florence, Italy.

PALAIS, J. M. & SIGURDSSON, H. 1989. Petrologic evidence of volatile emissions from major historic and prehistoric volcanic eruptions. *In:* BERGER, A., DICKINSON, R. E., & KIDSON, J. W. (eds) *Understanding climate change.* Geophysical Monograph 52. IUGG Volume 7, 31–53.

PÉTURSSON, G., PÁLSSON, P. & GEORGSSON, G. 1984. Um eituráhrif af völdum skaftárelda. *In:* GUNNLAUGASSON, G., GUOBERGSSON, G., PÓRARINSSON, S., RAFNSSON, S & EINARSSON, P. (eds) *Skaftáreldar.* Mál Og Menning, Rekyavik, 81–97.

PILCHER, J. A. & HALL, V. A. 1992. Towards a tephrochronology of the north of Ireland. *The Holocene*, **2**(3), 255–259.

——, —— & MCCORMAC, F. G. 1995. Dates of Holocene Icelandic volcanic eruptions from tephra layers in Irish peats. *The Holocene*, **5**(1), 103–110.

POLWHELE, REVD R. 1793. *The History of Devonshire.* (Vol I.) Exeter.

RENFREW, C. 1979. The eruption of Thera and Minoan Crete. *In:* SHEETS, P. D. & GRAYSON, D. K. (eds) *Volcanic Activity and Human Ecology.* Academic Press, New York and London, 565–582.

SIGURDSSON, H., DEVINE, J. D. & DAVIS, A. N. 1985. The petrologic estimate of volcanic degassing. *Jökull*, **35**, 1–8.

SIMKIN, T., SIEBERT, L., MCCLELLAND, L., BRIDGE, D., NEWHALL, C. & LATTER, J. H. 1981. *Volcanoes of the World.* Hutchinson Ross, Stroudsburg.

STOTHERS, R. B. 1996. The Great Dry Fog of 1783. *Climatic Change*, **32**(1), 79–89.

THÓRARINSSON, S. 1979. On the damage caused by volcanic eruptions with special reference to tephra and gases. *In:* SHEETS, P. D. & GRAYSON, D. K. (eds) *Volcanic Activity and Human Ecology.* Academic Press, New York and London, 125-59.

—— 1981. Greetings from Iceland: ash falls and volcanic aerosols in Scandinavia. *Geografiska Annaler*, **63A** (3-4), 109–18.

THORDARSON, TH. & SELF, S. 1993. The Laki (Skáftar Fires) and Grímsvötn eruptions in 1783-85. *Bulletin Volcanologique*, **55**, 233–263.

TOALDO, G. 1784. Dei principali accidenti dell'anno 1783. *Giornale Astrometeorologico per l'anno 1784.* Storti, Venice.

TORRENCE, R., SPECHT, J. & FULLAGAR, R. 1990. Pompeiis in the Pacific. *Australian Natural History*, **23** (6), 457–463.

WHITE, G. 1789. *The natural history of Selbourne* (reprinted 1977). Penguin, London.

WILES, R. M. 1965. *Freshest advices: early provincial newspapers in England.* Columbus, Ohio

Human adjustments and social vulnerability to volcanic hazards: the case of Furnas Volcano, São Miguel, Açores

DAVID K. CHESTER,[1] CHRISTOPHER DIBBEN,[2] RUI COUTINHO,[3] ANGUS M. DUNCAN,[2] PAUL D. COLE,[4] JOHN E. GUEST[4] & PETER J. BAXTER[5]

[1] *Department of Geography, University of Liverpool, Liverpool L69 3BX, UK*
[2] *Department of Geology, University of Luton, Luton LU1 3JU, UK*
[3] *Universidade dos Açores, Ponta Delgada, São Miguel, Açores, Portugal*
[4] *Planetary Image Centre, University College, University of London, London WC1E 6BT, UK*
[5] *Department of Community Medicine, University of Cambridge, Cambridge CB1 2ES, UK*

Abstract: Following a review of recent developments by social theorists, concerned with the policy and practice of natural hazard reduction it is argued that neither the so-called **dominant** nor the **radical** perspectives, are sufficient on their own to answer the questions now being addressed by applied volcanologists during the International Decade for Natural Disaster Reduction. In order to make useful recommendations to policy-makers, which are both reliable assessments of eruption risk and at the same time sensitive to probable interactions of future eruptions with the physical environment of the study region and its peoples, a conflation of traditional hazard analysis and vulnerability analysis is required.

This view is tested by means of a case study of evacuation planning at Furnas Volcano in the Açores. As well as supporting the argument from theory, this example leads to the proposition that research on evacuation planning and civil defence should proceed in an ordered sequence. First hazard analysis (i.e. hazard mapping) should be carried out in the traditional manner. Next possible interactions with the physical environment should be specified and a preliminary plan of evacuation routes drawn up. Before detailed recommendations about the logistics of evacuation are made, however, it is vital that such factors as the demographic, socio/economic and cultural/behavioural characteristics of the population at risk are fully considered.

In 1983 Kenneth Hewitt edited a volume, *Interpretations of Calamity*, which has subsequently become a 'classic' statement on the complexities inherent in natural hazard reduction (Hewitt 1983a). *Interpretations of Calamity* drew on ideas present, though inchoate, in the 1970s (e.g. Hewitt 1976, 1980; O'Keefe *et al.* 1976; Wisner *et al.* 1976, 1977) and represented a drawing together of disparate strands to produce a unified critique of the ways in which earth scientists, social scientists, national policy-makers and international agencies have traditionally attempted both to understand and react following extreme events of nature, so as to reduce mortality, morbidity and economic losses. Since 1983 further contributions have been published (eg. Whittow 1987; Palm 1990; Chester 1993; Blaikie et al. 1994) and today social theory – both traditional and radical – provides a rich fare of possible approaches from which the applied earth scientist may choose in making recommendations to policy makers about the measures required to reduce the hazardousness.

In this paper it is argued that when making recommendations to policy-makers about measures to reduce the impact of natural hazards, this rich theoretical fare should be seen not as a set of alternative courses, but rather as a smorgasbord from which elements may be chosen so that policy is sensitive not only to the volcanological circumstances of a particular site, but also to the complexities of the society in which it is located. These contentions are supported by two arguments: one from social theory, the other from current empirical research in the Açores. First it will be argued that both the traditional, or **dominant**, and the **radical** approaches are not sufficient on their own to capture fully the complexities involved in working at the interface between an erupting volcano and a vulnerable human population. A middle way between these approaches is proposed. Second, the proposed methodology is supported by detailed testing, by means of a case study of evacuation planning at Furnas Volcano, São Miguel, Açores.

Dominants and radicals: recent developments in social theory

Before the 1980s there was one approach to the study of natural hazards and their consequences which overshadowed all others. As might be expected this became known as the **dominant** approach and was first developed in the 1940s by Gilbert White when he was considering the range of possible policies which could be introduced to reduce flood losses in the USA (White 1942). Between the 1950s and late 1970s Gilbert White was joined by like-minded scholars, including Robert Kates and Ian Burton, and their approach was developed into a set of techniques which were used not only to study flooding in the USA, but also the totality of natural hazards occurring throughout the world. These scholars have achieved a commanding influence on the literature relating to hazard reduction. In turn this literature has largely controlled the ways in which policies of hazard reduction are framed by national governments, the United Nations Disaster Relief Office and, more recently, academics and policy-makers working within the context of the International Decade for Natural Disaster Reduction (IDNDR).

The characteristics of the dominant approach are complex in detail and further information may be gleaned from: Burton et al. (1978), Warrick (1979), Whittow (1987), Smith (1992) and Chester (1993). White and his co-workers accepted that such factors as material wealth, experience of hazardous events, systems of belief and psychological factors (e.g. Simon 1957, 1959) were all important in determining how individuals, social groups and societies responded to extreme events of nature, but at its heart their approach is based on the notion that there exists a range of so-called 'adjustments' to natural hazards, which are available to individuals and/or societies to deal with extreme events of nature. Table 1 is an example of adjustments currently available to deal with hazards from lava flows and similar tables could be constructed for other types of volcanic hazard. 'Bearing the loss' has been the involuntary adjustment which human societies have had to accept throughout most of their histories and this situation still occurs in poor countries today. In many pre-

Table 1. *The theoretical range of adjustments to hazards from lava flows*

			Adjust to losses		
Affect the cause	Modify the hazard	Modify the loss potential	Spread the losses	Plan for losses	Bear the losses
Types of adjustment No known way of altering the eruptive mechanism	(1) Protect high value installations	(1) Introduce warning systems	(1) public relief from national and local government	Individual family of company insurance	Individual family, company or community losses-sharing
	(2) Alter lava flow direction	(2) Prepare for a disaster through civil-defence measures	(2) Government sponsored and supported insurance schemes		
	(3) Arrest forward motion	(3) Introduce land-planning measures to control future development in particularly hazard-prone areas	(3) International relief from agencies such as the United Nations Disaster Relief Office		
Examples and notes	(a) Use of explosives and bombing to divert flows; has been tried in Hawaii and on Etna Emergency barriers tried in Hawaii, Japan and Etna Barriers to divert future flows from inhabited areas; have been suggested for the town of Hilo Hawaii) Control forward advance by watering the flow margin; limited success in Hawaii and Heimaey (Iceland) See references for	(b) Warning systems only available on certain well monitored volcanoes, in technologically advanced countries, e.g. USA (Hawaii and volcanoes showing signs of activity in the continental USA), Japan, Iceland and Italy Emergency evacuation plans have been formulated in several countries, e.g. USA, Japan and Soufriere de Guadeloupe Land-planning policies are in operation in some areas where 'general prediction' and hazard mapping have been carried out in the past, e.g. details	(c) Public relief available in most countries; the most comprehensive schemes are in the technologically most developed countries, e.g. Canada, USA Japan and New Zealand Government sponsored insurance schemes available in several countries, e.g. USSR and New Zealand UN Disaster Relief Office established only in 1972. May be of great benefit to developing countries in the face of major losses in the future; international relief given by many developed countries Paricutin eruption, Mexico	(d) Possible to a certain extent in more developed countries, but even in the USA it is limited by the discretion of individual companies	This is the traditional form of adjustment and is still widely practised in many volcanic areas

From Chester 1993, based on: Burton *et al.* 1978, Chester *et al.* 1985 and numerous other sources

industrial societies, however, more flexible adjustments showing greater harmonization with nature have always featured as a response (White 1973). For example, members of a pre-industrial group in Papua New Guinea, whose agriculture benefits from frequent pyroclastic falls, have developed rituals to encourage their recurrence (Blong 1982, 1984: 348).

Alternatives to 'bearing the loss' involve policy initiatives and over the last half century measures to: affect the cause of an extreme natural event; modify the hazard; modify its loss potential and adjust to losses (Table 1), have been introduced for a wide range of natural hazards in most rich countries of the world. Regardless of the type of hazard, policies designed within the framework of the dominant approach have been successful in reducing deaths and injuries and, although total economic losses have increased as a consequence of the process of development and wealth accumulation, the relative toll – as a percentage of national wealth – has fallen. For instance, the high magnitude eruption of Mount St. Helens in 1980 caused an estimated total loss of US$860 million (Blong 1984: 356), yet this only represented around 0.03% of the country's gross national product (GNP), while in Japan, across a range of natural disasters and with the exception of the recent Kobe earthquake, deaths normally represent only *c.* 3% of total casualties (Chester 1993: 241; Nakabayshi 1984).

In order to spread the benefits of these policies more widely national governments and international agencies have been and largely remain firmly wedded to the agenda set by the dominant paradigm. Although it is admitted that no hazard can exist unless there is a human population to be affected, the primary emphasis of the dominant approach is that physical processes – particularly the magnitude and frequency of extreme events – are 'first order determinants of a disaster and that differences between societies are at a lower, albeit still significant, level of importance' (Chester 1993: 237–238). Focus on the physically determined character of natural disasters is reinforced because research into hazard reduction has been traditionally under the control of scientists with backgrounds in geology, geophysics, and engineering. A physically deterministic locus may be perceived in the 'mission statements' of agencies in many countries which are charged with reducing the vulnerability of their populations to the effects of volcanic eruptions. This applies even if countries are at markedly different levels of economic development. The aims and objectives of the United States Volcanic Studies Program (Filson 1987: 294; Wright & Pierson 1992: 6), are remarkably similar to those of several Latin American countries (Zupla 1993).

A second emphasis is that the roles of national governments and international agencies are largely defined in terms of a transfer of the technology and administrative experience from areas where responses are observed to have been successful to those where they are either non-existent or are perceived to have failed. Internationally this means from rich countries to poor countries. This is exemplified by the implicit and explicit stress placed on technological transfer by some of the principal 'actors' involved in the IDNDR. For Professor Michel Lechet – a member of the United Nations IDNDR Scientific and Technical Committee (Lechet 1990: 1) – technology transfer is one of the principal objectives of the decade and for many – but not all – participants at recent international conferences (e.g. Anon 1992, 1993), it is a factor of overwhelming importance.

At first sight the acceptance of a model of loss reduction which has been successful in some countries as a universal international panacea may appear to be justified both intuitively and from a humanitarian perspective. This is, however, disputed by the **radical** critics. The research of radical scholars is heavily focused on poor countries and natural disasters, like droughts, which have a long onset time, are of long duration and cause damage to large areas. Although the arguments used by the radical critics are involved (see: Hewitt 1983a; Chester 1993; 238–44 for a fuller exposition), the idea of fundamental importance for writers like Kenneth Hewitt and Ben Wisner is that most disasters in poor countries have more to do with poverty and deprivation, than with the extreme character of physical processes. Kenneth Hewitt (1983b: 26), for instance, poses a question, 'what is more characteristic (for the inhabitants of the frequently drought ridden areas of the

Sahel) ... and to be expected by its long term inhabitants: recurrent droughts or the history of political, economic and social change'? Taking the analysis further Susman *et al.* (1983), made use of the Marxist notion of **marginalization**, by arguing that the people who suffer most in natural disasters are those who are either economically marginalized (i.e. poor) and/or geographically marginalized (i.e. live in areas which are prone to disaster losses). Relief aid and technological transfers have a tendency to benefit those who are already well off and can lead to further marginalization of the poorest sections of a community.

Some of the points raised by the more radical of the radical critics, such as Susman *et al.* (1983: 279–280), constitute an overtly Marxist agenda and because of this have not found favour with international agencies. It may be contented, however, that within the radical corpus there are themes which are of considerable value and should be taken on board by applied volcanologists. This may seem to be an eccentric proposition at first sight since volcano-related disasters are quite unlike droughts and many other classes of Third World hazard, because eruptions are usually marked by rapid onset, short duration and limited areal affects. As Russell Blong (1984) has shown, however, following many eruptions in economically less developed countries the success of recovery programmes depends critically upon the degree to which existing trends in development are either exacerbated or reversed. It is also the case that for eruptions of a similar magnitude, those occurring in poor countries not only exact high death tolls, but may also cause catastrophic economic losses. For example, in addition to a death toll of more than 25 000 people (Decker & Decker 1989), property losses following the 1985 Nevado del Ruiz eruption in Colombia were estimated by Sigurdsson & Carey (1986) at US$300 million, or *c.* 1% of GNP; with property losses representing only 10% of total economic losses. In fact the GNP 'costs' of a selection of natural disasters in the Third World are estimated at between 15 and 40% (Harriss *et al.* 1985; Kates 1987).

What is important about the radical critique, is that it emphasizes the uniqueness of place: Hawaii is not Etna and Iceland is not the Azores. Even when discussion is restricted to economically developed countries, successful hazard reduction depends critically on not only understanding volcanological processes *per se*, but also on the impacts these will have – on the wider physical environment and the fine detail of the socio/economic conditions and cultural milieux of the society in question. In short, responses and adjustments must be sensitive to the local environment and 'incultured' if they are to have any chance of success. This conclusion may be supported from the literature. One good example is Indonesia. Despite being a very poor country with a GNP per capita of only US$670 in 1992, Indonesia has responded exceptionally well to a range of volcano related emergencies in recent decades. This is due to a number of factors, of which the high priority given to applied volcanology in government spending and the important place of eruptions within the nation's history and cultural consciousness are of supreme importance (Zen 1983; Kusamadinata 1984). A second example is Italy. Italy is one of the twenty richest countries in the world (GNP per capita US$16 850 in 1992), yet until significant changes in policy were introduced in the middle 1980s its response to volcanic emergencies was more typical of a Third, rather than a First, World country. Again this was due to deep seated cultural and historic traits (Barberi *et al.* 1984; Chester 1993).

It is our contention that, whilst the dominant approach still provides a useful initial framework for translating volcanological research into policies aimed at loss reduction, it is equally vital that the uniqueness of place and the reasons why populations are vulnerable in these areas are not forgotten. Although some general principles of hazard reduction apply to all volcanoes regardless of location, the complex interactions between eruptions, environment, economy and society make all volcanic regions unique and demand an approach to hazard reduction which is of a more subtle character than has often been the case hitherto. Piers Blaikie and his colleagues have recently posited that what is required is an integration of conventional hazard analysis (i.e. mapping) with what they term **vulnerability analysis**. Whereas hazard mapping is concerned with physical mechanisms, vulnerability analysis concerns 'everything else: monitoring changes in root causes..., and understanding how these are channelled into unsafe conditions for specific subgroups in the population by social and economic mechanisms' (Blaikie *et al.* 1994: 225). Reworking their argument specifically in terms of volcanic hazards, it may be contended that integration involves a conflation of conventional volcanic hazard analysis through **general** and **specific** prediction (Walker 1974; Latter 1989), with study of those aspects of the human geography of a volcanic region which exacerbate, or diminish, its inhabitants susceptibility to losses. Some aspects may be independent of the volcanic threat (e.g. changes in population structure), being deep seated within the history of the region and/or reflecting dynamic changes in its culture and society, but other aspects of vulnerability are linked to eruptions and involve interactions between an erupting volcano and its region. These interactions not only concern the wider physical environment of the volcano, but also the communities which will be affected by an eruption and the behavioural characteristics of people living in them.

Evacuation planning at Furnas Volcano, São Miguel, Açores

In order to test the above contentions research was conducted on the issues involved in planning an evacuation of the region around Furnas Volcano (São Miguel, Açores – Fig. 1). This formed part of a wider international study of the pure and applied volcanology of Furnas.

Research into evacuation planning started with hazard analysis, to enable a future eruption scenario to be defined. Second, interactions with and the likely impacts of this scenario on the environment of the Furnas region were considered and this led to an analysis of possible evacuation routes. To this end the methodology is faithful to the dictates of the dominant approach, but it will be argued that it is only when such studies are considered along side information on demographic, socio/economic and cultural/behavioural aspects of the region, that an evacuation plan may be proposed which has a realistic chance of being successful.

Furnas Volcano: eruption history, conventional hazard mapping and the 1630 eruption scenario

Furnas Volcano is the most easterly of the three trachytic central volcanoes that are considered to be active on the island of São Miguel (Booth *et al.* 1978, 1983; Moore 1991*a*, 1991*b*). The others are Sete Cidades and Aqua de Pau i.e. Fogo (Fig. 1). Most of the eruptive history of Furnas has been explosive in character, with

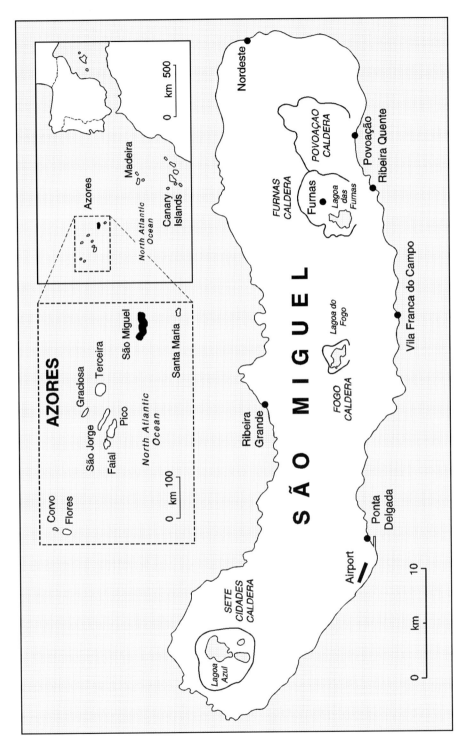

Fig. 1. Map showing the location of the Azores and the calderas of São Miguel.

Table 2. *A summary of the historic volcanic activity at Furnas Volcano (Guest et al. 1994)*

Point	Remarks
1	Over the past 5000 years eruptions have not been equally spaced in time.
2	Of known eruptions at least three occurred in the period from the beginning of the fifteenth century and up to and including AD1630, having a mean repose of only c. 77 years.
3	Before the fifteenth century there were six eruptions between 1000BC and AD1000, with a mean repose period of c. 330 years.
4	Between 1000 and 3000BC there is no evidence of intra-caldera explosive volcanic activity, though domes may have been formed and the region affected by ash-fall from eruptions of Fogo (i.e. Agua de Pau) Volcano (Booth et al. 1978).
5	The last eruption occurred in 1630.

eruptions often culminating in the formation of domes. The volcano is truncated by a caldera complex with steep walls (Fig. 1), and the formation of these calderas is probably associated with major explosive ignimbrite eruptions. Outside the caldera complex on the volcano's flanks are a few basaltic cinder-cones and lava flows, together with some trachytic centres.

The last eruption occurred in 1630, and during the last 5000 years stratigraphic studies show that the volcano has erupted at least 10 times (Booth et al. 1978; Guest et al. 1994). On the basis of this, Moore (1990) has argued that the volcano is long overdue for an eruption, not having erupted for three and a half centuries. The distribution of eruptions with time, however, has not been even and the volcano may be in a longer period of repose at present (Table 2).

Most known types of explosive volcanism have occurred at Furnas at some stage in its history (Guest et al. 1994), but during the last 5000 years typical eruptions have been plinian or sub-plinian in character, occurring within the caldera complex. In addition to magmatic explosions, interaction between ground water or lake water and magma has given rise to phreatomagmatic explosive activity.

Three eruptions have occurred on the island since the beginning of the fifteenth century, two of them since it was colonized in the mid fifteenth century (Queiroz et al. 1995). The most recent eruption, that of 1630, is well documented in contemporary accounts (Cole et al. 1995). Fortunately few people were living in Furnas at the time. Warning of the eruption was given by earthquakes that started up to 18 hours before the eruption began. These earthquakes destroyed the towns of Povoação and Ponta Garça (Fig. 3). The explosive phase of the eruption lasted three days and covered the surrounding area with ash-fall; pyroclastic surges devastating the area between the caldera and Ponta Garça killing 80 people. Pyroclastic flows also swept down the valleys of the area (Cole et al. 1995). Landslides occurred, as well as mudflows and water floods and even at Ponta Delgarda, over 30 km away, fine ash fell.

From hazard studies (Guest et al. 1994), it is predicted that with an eruption of the magnitude of 1630 the towns of Povoaçao and Ponta Garça (Fig. 3) would be covered with a maximum of 1 m of ash. In comparison with some eruptions during the last 5000 years, that of 1630 was relatively small. It probably represents the most common magnitude event to occur at the Furnas centre, and forms the starting point for evacuation planning and hazard mitigation.

Based on analysis of the 1630 eruptive products and historical accounts, the next

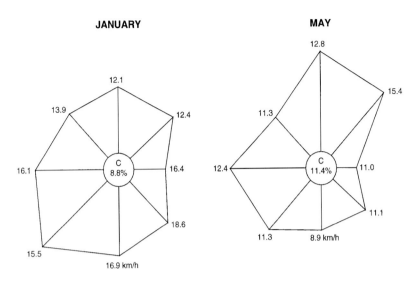
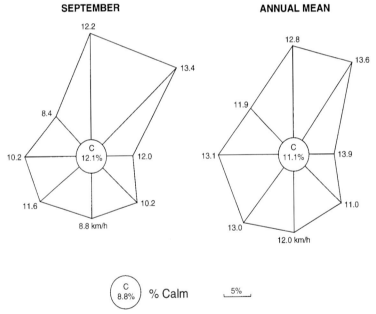

Fig. 2. Wind roses for Ponte Delgada meteorological station. 'C' represents calm conditions, the numbers around the edges average wind speeds from the direction shown.

eruption at Furnas will require rapid evacuation of the caldera area and surrounding towns. Precursory earthquakes could do serious damage to housing and roads. During the eruption everywhere within the caldera complex is likely to be devastated by pyroclastic surges together with thick ash and lapilli falls. Outside the caldera pyroclastic surges could reach towns such as Povoaçao and Ponta Garça, depending upon where the eruption takes place. The high walls of the caldera to the east and north may prevent surges spreading outside the caldera in these directions. Depending on wind direction, it is likely that almost the whole island will suffer some ash-fall. If the next eruption were larger than that of 1630, then ash-fall would be more extensive and problems for communications much greater.

The 1630 eruption scenario: impacts on the physical environment

Whereas conventional hazard analysis is able to define the areas which would be affected by eruption products and the historical record gives valuable clues to the probable course of a future eruption, it is clear that on a volcano like Furnas the hazardousness of a particular settlement and, consequently, constraints on any evacuation from it, have as much to do with likely interactions between an eruption and the external physical environment as they have with the eruption itself. Climatic and hydrological/geomorphological interactions have emerged as being particularly important.

Climatic factors. Although there are severe problems in using data from Ponta Delgada (Figs 1 and 2) – the only meteorological station on São Miguel with a comprehensive record – to assess conditions within the Furnas District, certain important features do emerge especially when data are combined with other published information more specific to the field area (i.e. Agostinho 1938–41; British Admiralty 1945; British Air Ministry 1949; Ferreira 1981*a,b*; Moreira 1987; SREA 1988; British Admiralty Pilot 67, 1992). First, unlike the situation in many parts of the world, climatic conditions in the Açores are not severe and in near sea-level parts of the Furnas District there is little chance of the people displaced by a volcanic eruption suffering the effects of exposure, as is frequently the case following eruptions in high latitudes and/or altitudes. In view of the height range found within São Miguel (Fig. 3) there is, however, a severe risk of exposure in the hills towards the centre of 'Furnas District', especially if an eruption were to occur in winter and/or at night.

Wind roses for Ponta Delgada are shown in Fig. 2. As far as evacuation is concerned the danger of pyroclastic fall materials covering a particular town or village during an eruption is the mirror image of the pattern shown by the wind roses. Hence, if an eruption where to occur in winter, then the more likely sectors to be affected by pyroclastic fall materials are the northeast, east and southeast, whereas in summer tephra fall is more likely to occur to the south and southwest of the caldera. In terms of evacuation planning the fine detail of Fig. 2 and other sources is more important than general trends, because in the Furnas District:

(a) winds from all directions are possible throughout the year;
(b) during the summer winds are light and calms are common;
(c) gales are common in winter;
(d) because of light pressure gradients and the considerable relief amplitude between the interior and coast of São Miguel (Fig. 3), katabatic ('drainage') winds (ie. coastal land breezes) are common on winter nights, to be superseded in the afternoons by sea breezes, particularly in summer.

Precipitation on São Miguel is high and can occur in all seasons. There is no dry month. Figures for Ponta Delgada show an average of 1313 mm, with some 64%

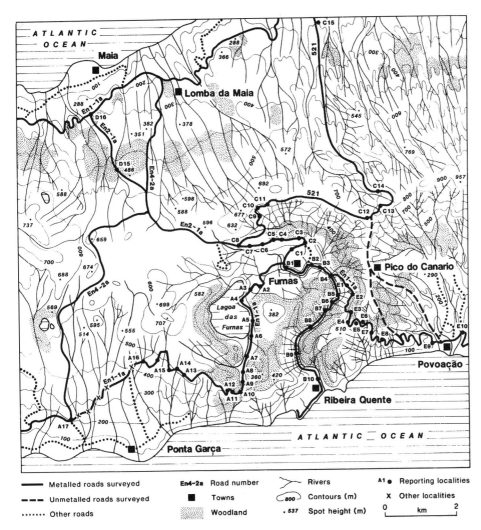

Fig. 3. Roads of the area around Furnas Volcano. The reporting localities represent points where conditions before and during an eruption could cause difficulties in the event of an evacuation.

falling between October and March inclusive. Heavy precipitation, most of it rain but with some hail and even snow at altitude in winter, is even higher away from the north and south coasts of the Furnas District and is frequently associated with mist, low cloud and poor visibility, particularly at high altitudes. Poor visibility even outside the area of an eruption plume would greatly hinder evacuation, as it does ordinary day-to-day communications at certain times each year (Stieglitz 1990), while high rainfall could pose the threat of tephra re-mobilization.

Hydrological/geomorphological factors. High rainfall forms a link between meteorology on the one hand and geomorphology, hydrology and soil stability on the other. São Miguel has a dense drainage pattern (Fig. 3) and a history of

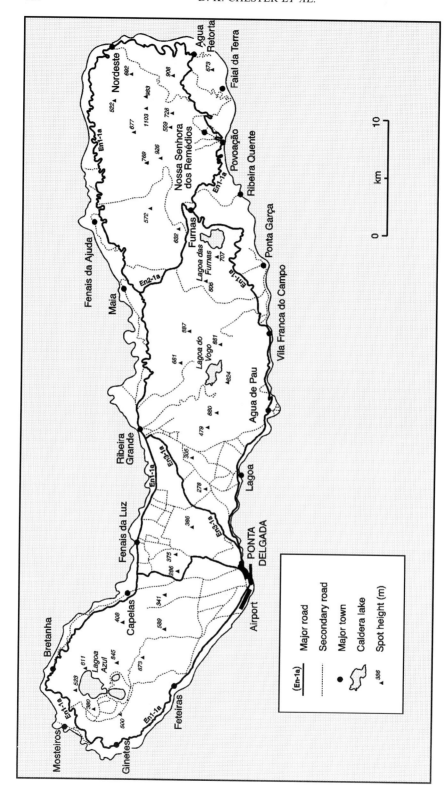

Fig. 4. Principal roads of São Miguel.

Table 3. *Major constraints on the use of evacuation routes from the Furnas District*

	Major constraints
1.	The major constraint involves removing people quickly from the caldera region. In addition to the hazards from primary volcanic processes discussed in the text, there are major flood, debris flow and landslide risks on the road running by the side of the Lagoa das Furnas (Fig. 3). Route 521 (reporting localities C8–12) runs along the rim of the caldera for a considerable distance and is potentially very dangerous and the best overall route for evacuation is the EN2-1a, which runs to the west and northwest of Furnas. Even on this route, however, there are some serious potential problems between localities C1 and C5, due to a potential lack of carriageway stability and possible landsliding. The EN2-1a once it emerges from the caldera at locality C8, is an excellent route to the west of the island.
2.	Ribeira Quente emerges as a very dangerous settlement. Not only would flood waters be concentrated within the valley leading to the village (see text) but, between reporting localities B4 and B10, the road has several major physical constraints on its use during an eruption. These including hazards from falling trees, landslides, flooding and carriageway instability. It is likely that once an eruption started this route would be unusable and seaborne evacuation should have to be contemplated.
3.	Evacuation from other settlements shown on Fig. 3 does not seem to be too difficult. Evacuation from Povoaçao and Pico do Canario would, however, require a long detour around the east of the island.

This table should be read in conjunction with Fig. 3.

landslides, many of which have blocked roads. Both these factors could affect the potential use of an evacuation route. In winter, high rainfall in the period before an eruption would not only increase landslide risk, but could also lead to a reduction in intra-basin storage capacity, an increase in flood stage height along the rivers and streams of the region and a decrease in soil storage capacity. The effects of any increase in discharge in the early stages of an eruption due, say, to breaching of the caldera wall or a landslide into the Lagoa das Furnas would greatly exacerbate flooding. Flooding risk would, moreover, be increased if the **Factor of Safety** (Innes 1983) of the often poorly consolidated wall rocks of the Lagoa das Furnas were to be exceeded, so that a large volume of water drained into the basin of the Ribeira Almarela which flows through Furnas village. The Ribeira Almarela is the principal tributary of the Ribeira Quente which passes close to the village with the same name (Fig. 3). Considerable devastation in the village of Ribeira Quente and the coastal zone around it is, thus, highly probable and a special effort would have to be made to evacuate this village and its environs. Flooding could also occur if an eruption occurred on, or close to, the caldera wall so breaching it. Hence, especially in winter, planning for evacuation requires a detailed knowledge of previous rainfall conditions, their effects on soil moisture and the hydrological system.

On the basis of hazard analysis and study of likely interactions between a possible future eruption and local climate, hydrological and geomorphological factors, field work on evacuation routes was carried out (Figs 3 & 4). As Fig. 3 shows all routes leading from Furnas village were surveyed and each 'reporting locality' represents a physical and logistical constraint on the use of the road at or near to the locality in question. These constraints, discussed in detail by Chester *et al.* (1995), are summarized in Table 3.

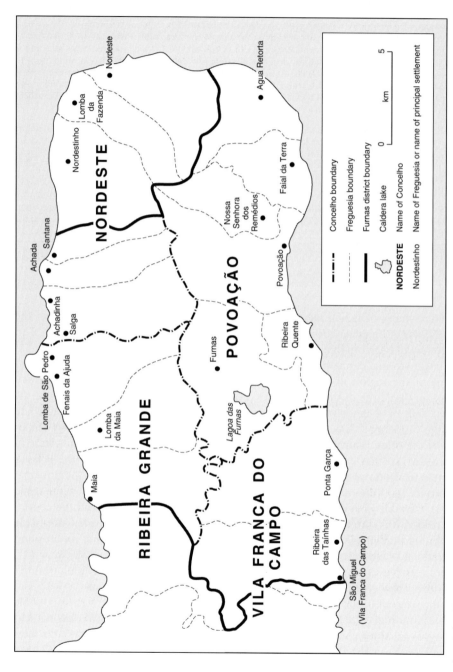

Fig. 5. Limits of the 'Furnas District', showing administrative boundaries. The administrative boundaries correspond to subdivisions used for official statistics purposes.

Table 4. *A summary of the main demographic, socio/economic and cultural/behavioural factors which could complicate an evacuation*

Factors	Details
Demographic	The *Furnas District* (Fig. 5) had a resident population in 1991 of 22 644 people, accommodated in 5693 houses. These figures do not capture: (a) the large number of visitors, especially in summer and (b) the under-occupancy of many houses for most of the year. These are inhabited for some of the time by weekend residents, return migrants from mainland Portugal and abroad and by tourists. Since the Portuguese census date is 15 April, a month of relatively low resident population, far more people would potentially require transporting to safety at certain times of the year than the published figures suggest. A further demographic factor which could complicate evacuation is the degree to which population is either clustered within the settlement, which gives it name to the *concelho/freguesia*, or widely dispersed over its area.
Socio/economic	Traditionally an agricultural/fishing area, alternative employment opportunities are limited. This is reflected in recent decades by the permanent or temporary out-migration of many people in the economically active cohort, with the result that 'dependency ratios' (i.e. % under 15 and % over 65 in the population) range from 38–46% across the district. The percentage of the population classed as economically active is never greater than 36% of the total in any *freguesia*.
Cultural/Behavioural	As a result of an aged population and the fact that before the 1974 Portuguese Revolution many people did not receive even an elementary education, illiteracy ranges from 8–23% of the total population, across the Furnas District. Also, behavioural factors are important and include close links between many of the inhabitants and the land, as a result of both active and traditional family based agricultural ties. At its most simple and bearing in mind that cattle rearing and fattening dominate contemporary agriculture, large numbers of livestock – both living and dead – could block roads in the event of an eruption, whilst the close attachment of people to village, land and farm could cause some inhabitants to resist evacuation.

Based on information in Chapin (1989), Correa (1924), Fortuna 1988, SREA 1993*a, b*, interviews carried out in the field and the authors' field work.

The 1630 eruption scenario: economic, social and behavioural considerations

It might be thought that, although time consuming, moving from what has been outlined above to a fully developed evacuation plan is simply a question of logistical planning. In other words the task is merely to integrate hazard analyses and any environmental constraints placed upon them, with plans based on such considerations as: the population of each settlement; inventories of refugee accommodation; vehicle availability and hospital beds. Indeed such studies are being carried out as part of the CEC Furnas Project and there are numerous precedents for research of this kind (e.g. Blong 1984; Zelinski & Kosiski 1991). This essentially managerial approach, however, carries the danger that the local people, for whom the whole exercise is being carried out, may be forgotten in the rush to publish a plan.

Following study of published materials on the demographic, socio/economic and cultural/behavioural characteristics of the communities which comprise the Furnas district, it became apparent that before any firm recommendations could be made to policy-makers about the logistics of evacuation there was a need to consider a range of non-volcanological and non-environmental factors. This impression was supported by interviews with local administrators, politicians and field work in the Furnas region. An area was defined 'Furnas District' (Fig. 5), which comprises *concelho* (i.e. municipalities) and *freguesia* (i.e. parishes) for which statistical data are published. The Furnas District corresponds to the administrative areas of the towns and villages which would be at risk if an eruption typified by the 1630 scenario were to recur.

Some of the major demographic, socio/economic and cultural/behavioural features of the Furnas District are listed in Table 4. Failure to be aware of the local milieu could seriously damage the effectiveness of any planned evacuation, and relevant aspects of society and culture need to be brought to the attention of policy-makers. Four characteristics in particular require special emphasis. First, because of a large transient population and the consequent inflation of population numbers particularly in summer, it was strongly recommended that policy-makers commission further research on the maximum number of people requiring evacuation at different times of the year. A second recommendation was also concerned with demography. Although population is concentrated in villages, special attention needs to be given to people living in isolated dwellings. These people should be located and detailed plans devised to ensure their safe evacuation. Third, because of high illiteracy rates and dependency ratios, many people are very young, very old and/or not able to follow written instructions. As a consequence many people will require considerable assistance in the event of an evacuation. Finally farmers have a strong attachment to their land, livestock and village. Great care will have to be taken to ensure that this social group is made aware that evacuation is necessary and that their livestock are properly controlled. As argued more fully in Chester et al. 1995: 21–26), identifying local leadership is vital if these and other sections of the community are not to be marginalized in any evacuation.

Conclusions

Steering a middle course between claims of dominant and radical social theorists, not only allows an approach which is faithful to the strengths of both traditions but also, through integrating hazard with vulnerability analysis, defines a methodology which is of practical use for studying the complexities inherent in making recommendations about loss reduction to policy-makers. It is contended that such a methodology can be applied to all countries – regardless of their level of economic development and not merely to those located at opposite ends of the continuum of economic development, as is the case, respectively, for the dominant and radical approaches.

Study of evacuation planning in the Azores not only supports these propositions, but also allows a sequential plan of action to emerge. This proceeds from: general and specific prediction; through study of possible interactions between an erupting volcano and its physical environment to considerations of the likely impacts upon communities. Only when all these data are collected and their interactions analysed may questions about logistics be considered. Research in the Açores also brings two additional issues into focus. First possible interactions between a volcano and its region are so potentially complex – and so ill-constrained are some of the important causal variables (e.g. weather, pre-existing hydrological conditions, population actually present and aspects of human behaviour) – that any plans must be flexible and capable of change. This applies both in the days and months leading up to an eruption and during its early stages. Information flow to local scientists and policy-makers is, therefore, vital and study of communication links between policy-makers/scientists and local communities is also being researched as part of the Furnas

Project. Second, because of the complexity of the issues an interdisciplinary approach is required. If possible this should be international, so that the experience of researchers who have worked on a number of volcanic regions may be combined with that which can only come from long acquaintaceship with a particular volcanic region and cultural milieu.

In summary, both recent developments in social theory and the empirical research reported in this paper, lend support to a view that uniqueness of place must be kept constantly in mind by the applied volcanologist.

References

AGOSTINHO, A. T. 1938–41. Clima dos Açores. *Açoreana, Agra do Heroismo*, **2**, 35–224.
ANON 1992. *Opportunities for British involvement in the International Decade for Natural Disaster Reduction (IDNDR): Proceedings of a workshop held on 27 March 1992*. The Royal Society and The Royal Society of Engineering, London.
—— 1993. *Medicine in the International Decade for Natural Disaster Reduction (IDNDR): Proceedings of a workshop at The Royal Society, London*. The Royal Society and The Royal Society of Engineering, London.
BARBERI, F., CORRADO, G, INNOCENTI, F. & LUONGO, G. 1984. Phlegraean Fields 1982–1984: Brief chronicle of a volcanic emergency in a densely populated area. *Bulletin Volcanologique* **47**, 175–185.
BLAIKIE, P., CANNON, T., DAVIS, I. & WISNER, B. 1994. *At risk: Natural hazards, people's vulnerability and disasters*. Routledge, London.
BLONG, R. J. 1982. *The time of darkness, local legends and volcanic reality in Papua New Guinea*. University of Washington Press and Australian National University Press, Seattle and Camberra.
—— 1984. *Volcanic hazards: A sourcebook on the effects of eruptions*. Academic Press, Sydney.
BOOTH, B., CROASDALE, R. & WALKER, G. P. L. 1983. Volcanic hazard on Sao Miguel, Azores. *In*: TAZIEFF, H. & SABROUX, J-C. (eds) *Forecasting Volcanic Events*. Elsevier, Amsterdam, 99–109
——, WALKER, G. P. L. & CROASDALE, R. 1978. A quantitative study of five thousand years of volcanism on Sao Miguel, Azores. *Philosophical Transactions Royal Society of London*, **A228**, 271–319.
BRITISH ADMIRALTY 1945. *Spain and Portugal: Volume IV The Atlantic Islands*. Naval Intelligence Division, London.
BRITISH AIR MINISTRY 1949. *Aviation meteorology of the Azores*. Air Ministry Meteorological Office, Meteorological Reports No. 2, London.
BRITISH ADMIRALTY PILOT 67, 1992. *West coasts of Spain and Portugal Pilot*. HMSO, Taunton.
BURTON, I., KATES, R. W. & WHITE, G. 1978. *The environment as hazard*. Oxford University Press, New York.
CHAPIN, F. W. 1989. *Tides of migration: A study of migration decision making and social progress in Sao Miguel*. AMS Press, New York.
CHESTER, D. K. 1993. *Volcanoes and Society*. Edward Arnold, London.
——, DIBBEN, C. & COUTINHO, R. 1995. *Report on the evacuation of the Furnas District in the event of a future eruption*. Open File Report, CEC Environment: ESF Laboratory Volcano, Furnas, Açores, Planetary Image Centre, University College, London.
——, DUNCAN, A. M., GUEST, J. E. & KILBURN, C. R. J. 1985. *Mount Etna: The anatomy of a volcano*. Chapman and Hall, London.
COLE, P. D., QUEIROZ, G., WALLENSTEIN, N. GASPAR, J. L., DUNCAN, A. M. & GUEST, J. E. 1995. An historic subplinian/phreatomagmatic eruption: The 1630AD eruption of Furnas Volcano Sao Miguel, Azores. *Journal of Volcanology and Geothermal Research*, **69**, 117–135.
CORREA, M. DE J. 1924. *Leituras sobre a Historia do Valle das Furnas*. Officina de Artes Braficas, S. Miguel.

DECKER, R. W. & DECKER, B. 1989. *Volcanoes*. W.H. Freeman, New York.

FERREIRA, D. DE B. 1981a. Les types de temps de saisons chaude aux Açores. *Finisterra*, **16**, 231–260.

—— 1981b. Les mécanismes des pluies et les types de temps de saison Fraîche sux Açores. *Finisterra*, **16**, 15–61.

FILSON, J. R. 1987. Geological hazards: Programs and research in the USA. *Episodes*, **10**, 292–295.

FORTUNA, M. J. A. 1988. A populaçao activa dos Açores e a sua distribuiçao sectorial. *Arquipélago (Revista da Universidade dos Açores, Economia)*, **1**, 41–59.

GUEST, J. E., GASPAR, J. L., DUNCAN, A. M., QUEIROZ, G, COLE, P. D., WALLENSTEIN, N. & FERREIRA, T. 1994. *Preliminary report on the volcanic geology of Furnas Volcano, Sao Miguel, The Azores*. Open File Report 1, CEC Environment: ESF Laboratory Volcano, Furnas, Açores, Planetary Image Centre, University College, London.

HARRISS, R. W., HOHENEMSER, C. & KATES, R. W. 1985. Human and non-human mortality. *In:* KATES, R. W., HOHENEMSER, C. & KASPERSON, J. X. (eds) *Perilous progress: Managing the hazards of technology*, Westview Press, Boulder, 129–155.

HEWITT, K. 1976. Earthquake hazards in the mountains. *Natural Hazards*, **85**, 30–37.

—— 1980. Review: 'the environment as hazard'. *Annals Association of American Geographers*, **70**, 306–311.

—— (ed.) 1983a. *Interpretations of Calamity*. Allen and Unwin, London.

—— 1983b. The idea of calamity in a technocratic age. *In:* HEWITT, K. (ed.) *Interpretations of Calamity*. Allen and Unwin, London, 3–30.

INNES, J. L. 1983. Debris flows. *Progress in Physical Geography*, **7**, 469–502.

KATES, R. W. 1987. Hazard assessment and management. *In:* MCLAREN, D. J. & SKINNER, B. J. (eds.) *Resources and world development*. John Wiley, Chichester, 741–753.

KUSAMADINATA, K. 1984. Indonesia. *In:* CRANDELL, D. R., BOOTH, R., KASUMADINATA, K., SHIMOZURU, D., WALKER, G. P. L. & WESTERCAMP, D. (eds) *Source-book for volcanic hazards zonation*. UNESCO, Paris, 55–60.

LATTER, J. H. (ed.) 1989. *Volcanic hazards, assessment and monitoring*. IAVCEI Proceedings in Volcanology, Springer-Verlag, Berlin.

LECHAT, M. F. 1990. The International Decade for Natural Disaster Reduction. *Disasters*, **14**, 1–6.

MOORE, R. B. 1990. Volcanic geology and eruption frequency, São Miguel, Azores. *Bulletin of Volcanology*, **52**, 602–614.

—— 1991a. Geology of three late Quaternary stratovolcanoes on Sao Miguel, Azores. *United States Geological Survey Bulletin*, 1900.

—— 1991b. *Geologic map of Sao Miguel, Azores*. Miscellaneous Investigation Survey Map 1-2007, scale 1 : 50 000 Washington DC.

MOREIRA 1987. *Alguns aspectos de intervençao humana na evoluçao da paisagem da Ilha de S. Miguel Açores*. Serviço Nacional de Parques, Reservas e Conservaçao da Natureza, Lisboa.

NAKABAYSHI, I. 1984. Assessing intensity of damage of natural disasters in Japan. *Ekistics*, **308**, 432–428.

O'KEEFE, P., WESTGATE, K. & WISNER, B. 1976. Taking the naturalness out of natural disaster. *Nature*, **260**, 566–567.

PALM, R. I. 1990. *Natural Hazards: An integrative framework for research and planning*. Johns Hopkins Press, Baltimore.

QUEIROZ, G., GASPAR, J. L., COLE, P. D., GUEST, J. E., WALLENSTEIN, N., DUNCAN, A. M. & PACHECO, J. 1995. Erupçoes volcanicas no Vale das Furnas (ilha de Sao Miguel) na primera metade do seculo XV. *Açoreana*, **8**(1), 159–165.

SIGURDSSON, H. & CAREY, S. 1986. Volcanic disasters in Latin America and the 13th November 1985 eruption of Nevado del Ruiz volcano on Colombia. *Disasters*, **10**, 205–216.

SIMON, H. A. 1957. *Administrative Behaviour*. Macmillan, New York.

—— 1959. Theories of decision making in economic and behavioral science. *American Economic Review*, **49**, 253–83.

SMITH, K. 1992. *Environmental hazards*. Routledge, London.

SREA 1988. *Anuário estatístico 1986/7 Açores*. Serviço Regional de Estatística dos Açores, Açores.
—— 1993a. *XIII Recenseamento geral da População, III Recenseamento geral da Habitaçao 1991*. Serviço Regional de Estatística dos Açores, Açores.
—— 1993b. *Turismo: Novembro 1993*. Serviço Regional de Estatística dos Açores, Açores.
STIEGLITZ, A. 1990. *Landscapes of the Azores (São Miguel): A countryside Guide*. Sunflower Books, London.
SUSMAN, P., O'KEEFE, P. & WISNER, B. 1983. Global disasters, a radical interpretation. *In:* HEWITT, K. (ed.) *Interpretations of Calamity*. Allen and Unwin, London, 263–280.
WALKER, G. P. L. 1974. Volcanic hazards and the prediction of volcanic eruptions. *In:* FUNNELL, B. M. (ed.) *Prediction of geological hazards*, Miscellaneous Paper 3, Geological Society, London, 23–41.
WARRICK, R. A. 1979. Volcanoes as hazard: An overview. *In:* SHEETS, P. D. & GRAYSON, D. K. (eds) *Volcanic activity and Human Ecology*. Academic Press, New York, 161–189.
WHITE, G. F. 1942. *Human adjustment to floods: A geographical approach to the flood problem in the United States*. Research Paper 29, Department of Geography, University of Chicago.
—— 1973. Natural hazards research. *In:* CHORLEY, R. J. (ed.) *Directions in Geography*. Methuen, London, 193–212.
WHITTOW, J. 1987. Hazard-adjustment and mitigation. *In:* CLARK, M. J., GREGORY, K. J. & GURNELL, A. M. (eds) *Horizons in Physical Geography*. Macmillan, London, 307–319.
WISNER, B., O'KEEFE, P. & WESTGATE, K. 1977. Global systems and local disasters: the untapped power of people's science. *Disasters*, **1**, 47–57.
——, WESTGATE, K. & O'KEEFE, P. 1976. Poverty and disaster. *New Society*, **9**, 547–8.
WRIGHT, T. L. & PIERSON, T. C. 1992. *Living with volcanoes: The US Geological Survey's volcanic hazards program*. United States Geological Survey, Circular, 1073.
ZELINSKI, W. & KOSISKI, L. A. 1991. *The emergency evacuation of cities*. Rowman and Littlefield, Maryland.
ZEN, M. T. 1983. Mitigating volcanic disasters in Indonesia. *In:* TAZIEFF, H. & SABROUX, J. C. *Forecasting volcanic events*. Elsevier Developments in Volcanology, Amsterdam, 219–236.
ZUPLA, D. 1993. UN/DHA action for disaster mitigation in Ecuador. *Stop Disasters*, **16**, 16–17.

Reconstruction of the 1706 Montaña Negra eruption. Emergency procedures for Garachico and El Tanque, Tenerife, Canary Islands

M. C. SOLANA & A. APARICIO

Dept. de Volcanología. MNCN-CSIC, C/ José Gutierrez Abascal, 2, 28006 Madrid, Spain

Abstract: In May 1706, after several earthquakes, an eruptive fissure opened in the hills of the Garachico district, forming the Montaña Negra Volcano. In <15 h the village of El Tanque and part of the town of Garachico, including the harbour, were buried under lava flows. A week later, another flow buried the village of San Juan del Reparo and much of the remaining town of Garachico.

Today, over 6700 people live in the immediate vicinity of this former eruption. Although most of the population is concentrated in the main villages, dwellings are located throughout the district and any future eruption will inevitably destroy property. The reconstruction of the 1706 eruption shows that in any future event only a short time will be available to evacuate the population, but it may be possible to try to prevent the effects of the flows.

The Canary Islands, located off the western continental margin of North Africa, are the only active volcanic zone in Spain. The islands have experienced 15 eruptions since AD 1500 (Fig. 1), the most recent being at Chinyero (1909) in Tenerife, and San Juan (1949) and Teneguía (1971) in La Palma. Although some of these eruptions may not have produced any direct casualties, the economic effects to the affected areas have been important, e.g. in the case of the Timanfaya eruption, 22% of Lanzarote was covered; whilst in the Garachico eruption one of the most important harbours of the Canaries was destroyed.

Although each eruption has been distinctive in its character, nearly all of them display a series of common features (Table 1). In general, the eruptions are effusive basaltic events, that extrude through fissures of variable lengths (metres to kilometres) with a strombolian character. The eruptions tend to last a few weeks to months and are associated with a relatively low volume of total emitted material. Each eruption tends to be preceded by local earthquakes (although some are felt on other islands) which may destroy houses, and produce rockfalls and, sometimes, visible deformation of the ground.

Reconstructing the events of past eruptions and comparing them with present conditions of land use and population density can certainly help in the evaluation of future risks, and this process has been developed by scientists in collaboration with civil protection officials to assess the measures that can and should be taken in the event of the next eruption.

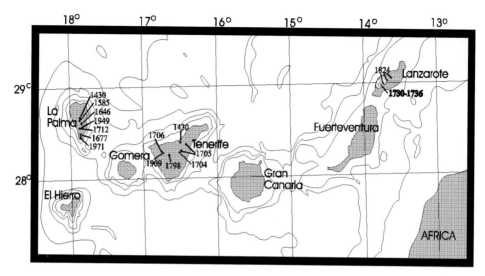

Fig. 1. Location of the historical eruptions of the Canary Islands.

Table 1. *General characteristics of Canarian historical volcanism (Araña, 1991)*

Volcano type	Duration (area covered by lavas)	Precursors	Location
Basaltic Monogenetic Strombolian cones Lava flows Vents lined up along tectovolcanic fissures	Weeks/Months (3–7 Km^2)	Seismicity (days/months before)	Ridges (volcano–Tectonic axis) and Holocene volcanic fields in all the islands
Exceptions Teide stratovolcano Freatomagmatic eruptions	Timanfaya: 6 years (200 Km^2)	Not always Deformation in 1585 eruption in la Palma	Gomera: no recent activity

The 1706 eruption

The 1706 Montaña Negra eruption began at 3.30 a.m. on May 5th, following a preliminary phase of earthquakes [the historical data concerning this eruption are reported in Romero (1991)]. Even though the historical references do not mention the activity on the vent, lava effusion must have begun almost immediately after the 850 m long fissure (Solana 1998) had opened at an altitude of 1300 m. The lava flowed from several boccas aligned along the fracture but was concentrated in one vent at the highest point of the fracture.

By referring to the height of recent eruption columns of the volcanoes of 1909 ($\gg$600 m; Ponte & Cologán 1911), 1949 ($\gg$2000 m; Romero & Bonelli 1951) and in 1971 ($\gg$1500 m), which displayed similar periods of activity and cone heights, the 1706 eruption column has been estimated to be c. 1.5 km. The pyroclasts spread south–southwest for no more than 3000 m, in exactly the opposite direction to the

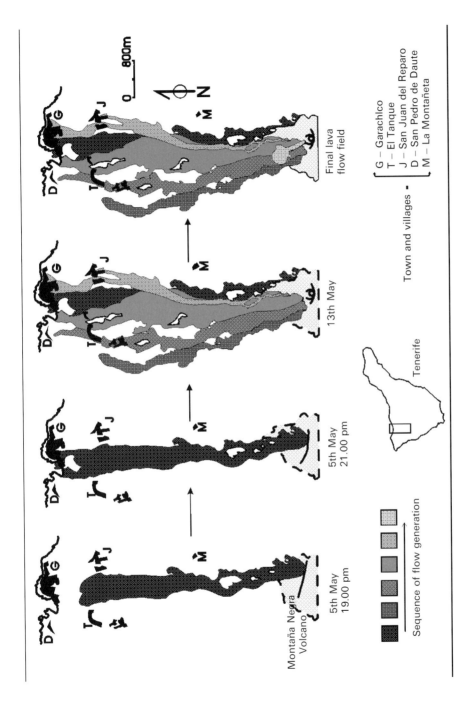

Fig. 2. Evolution of the lava flow field of Montaña Negra Volcano and the sequence of village invasion.

movement of the lavas.

Petrologically the lavas are basalts, basanites and phonolitic tephrites which advanced 7000 m over an 8° slope at a mean rate of $0.13\,\mathrm{m\,s^{-1}}$ for the first 15 h. The lavas reached the cliff above Garachico by 7 pm, having already destroyed the main part of the inland village of El Tanque. The lava descended the cliff in two main streams which turned and approached each other at the base, burying the western outskirts of Garachico and overwhelming two thirds of the harbour (Fig. 2). The lava flows stopped at Garachico soon afterwards, although activity must have continued at the vent, constructing a 120 m high strombolian cone at the head of a growing lava field.

Eight days later a new and more violent flow moved over and covered the village of San Juan del Reparo just before the lavas, confined in a ravine, entered Garachico, destroying most of the town, even though it did not reach the sea (Fig. 2). This new flow resulted from a break produced by the accumulation of lava in the eastern flank of the main flow (Solana 1998) and, geochemically, this last stage coincided with a loss of SiO_2.

This eruption resulted in the decline of Garachico, which was one of the largest and richest towns in the Canaries at that time. The town was the site of the island's most important harbour which was destroyed as a result of the eruption. Garachico, like many other villages of northwest Tenerife, was built on a pre-existing platform of recent (Holocene) lavas which developed at the foot of the palaeocliff that surrounds this side of the island. Even though the villages and the town of Garachico were rebuilt, the damage to the harbour was irreparable and the town declined from the island's major commercial centre to a fishing village.

Risk assessment and evacuation procedures

The risk of an area in terms of a potential hazard can be evaluated by studying the value and vulnerability of the area in particular as well as the hazard in question (UNESCO 1972). By using such analysis, three approaches have been identified for reducing the risk from a future eruption (Ayala-Carcedo 1993): (1) decreasing the vulnerable population; (2) installing protective measures such as lava diversion barriers or dams (Alexander 1993; Chester 1993); (3) improving emergency evacuation procedures. In the case of Tenerife, and especially Garachico, the first option is difficult to implement since the available habitable land is very restricted. Similarly, the second option is unattractive since the flat topography above the sea cliffs provides no natural basin for damming a flow, and the widespread distribution of the possible vents and of the local population yields no obvious location towards which lava can be diverted. As a result, a comprehensive evacuation plan is the primary objective for strategies to reduce the risk from lava flows in this area.

Some of the problems that arise when planning volcanic risk procedures in the Canary Islands stem from the fact that technological hazards, or even natural hazards (e.g. forest fires, flash floods or coastal flooding), are seen by the authorities as being of more immediate risk than earthquakes or volcanoes, which subsequently lose priority. Another problem is that classical volcanic hazard assessments (e.g. Booth 1979; Carracedo et al. 1990) do not help very much in the prevention and planning of the phenomena in detail (Araña 1989; Araña & Ortiz 1993).

The situation has, however, been clarified to some extent since February 1996,

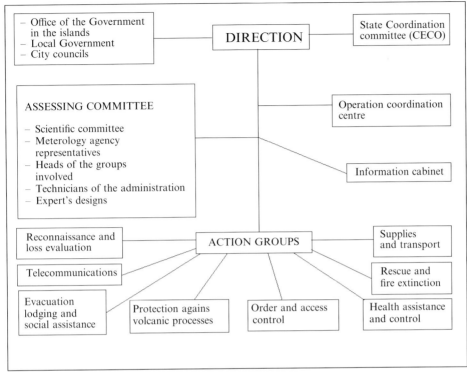

Fig. 3. Canary Island organizational structure in case of a volcanic crisis (Sansón 1996).

when a Spanish law established the minimum contents that both the Canarian and State plans for volcanic eruptions have to contain. The structure and organization at the Canarian level is shown in Fig. 3.

Potential risks and protection of properties if the 1706 eruption were to occur today

The time available in such situations condition the priorities adopted in evacuation procedures. In the case of a Garachico-type eruption, once the eruption starts the priority would be to evacuate people. The 10 h available to evacuate Cruz Grande, El Tanque and San Juan del Reparo and the 15 h available in Garachico, with only one road in each case (C-820 and Tf-142, respectively), does not give sufficient time to try to protect property and even less a harbour at the foot of a cliff. Any pre-eruption measure, such as barriers near towns, would be difficult to implement given the widespread distribution of the local population, the topography (relatively flat above the cliff approaching 8°, but lined with ravines which are very important for drainage) and the uncertain nature of where the eruption will occur.

Once the vent(s) has been located, the points at risk can be ascertained. In order to do this, the steepness of the slope that can favour the formation of channels that produce quicker displacement of the lavas, and the fact that these flows are not wider than 1000 m needs to be taken into account. Therefore, all the villages along a band with a width of the effective length of the fracture plus the same distance

Table 2. *Estimated arrival time to the main towns/villages for the lava flows in the 1706 eruption*

Town/village	Distance to crater (m)	Time of arrival (h)
La Montañeta	2423	6
San Juan del Reparo	4950	12
El Tanque	4090	10
Garachico	6160	15

(850 m in this case) on each side from the vents to the sea will be at risk. The situation would be: La Montañeta, El Tanque, San Juan del Reparo, San Pedro de Daute, Cruz Grande and El Guincho villages, and Garachico town would most probably be affected by lava invasion. Even though tunnels are not referred to in the 1706 eruption, they are common in Canarian eruptions and could produce an even faster displacement of the lavas.

Generally, the time between the opening of the fracture and the flow of the lava oscillates between minutes and hours. In this case, if the flow is considered to have started when the fissure opened, and assuming a regular rate of movement down an 8° slope (0.13 m s^{-1}, the average estimated from the historical records), the theoretical time of arrival at each village is given in Table 2. The evacuation plans and routes thus need to be able to move almost 7000 people to safety in the times given previously. This is possible with a good plan and efficient communication to previously informed and prepared groups. It is also important that the general population have been previously informed of basic procedures.

The velocity of the lavas is partly conditioned by their viscosity. In Canarian eruptions, changes in viscosity are frequent; e.g. in the 1971 eruption, lavas were more viscous in the first stages than in the later ones. This factor is very important if an estimate of the effective time for each operation is to be determined.

Once the main flow has developed, zones of lava accumulation and possible ruptures on the sides of the channels need to be monitored and protective measures taken where necessary, including the creation of openings with explosives to allow lava to discharge over previously covered zones. In the case of ruptures, areas may be protected by constructing rock barriers, which have proven their effectiveness in diverting flows – in general, a 3 m thick wall is enough to deviate a 1 m deep lava flow (Alexander 1993). In the case of San Juan del Reparo, the residents had a week before the flow buried the town. In such a case it would be possible to construct a barrier that could deviate the lava into the sea, away from Garachico.

The historical reports indicate that a series of strong ground movements preceded the 1706 eruption. The effects included not just vibrations of the ground but also faulting and rockfalls. Today, such physical ground movements could affect houses and damage structures such as bridges, tunnels and gas stations. Rockfalls or landslides could affect buildings and roads over a large area since gradients are steep. The population of El Guincho, Garachico and San Pedro de Daute, and the road that joins them (Tf-142), would be particularly at risk since they lie at the foot of a steep cliff.

Another possible hazard associated with Canarian eruptions is the propagation of the fracture (10 000 m in the case of the 1704–1705 eruption in Tenerife), thereby producing other vents. As a result, villages in line with the direction of the fracture should be prepared for this outcome.

In terms of hazards posed by ballistic projectiles, the risk is minimal since the potential site of the eruption is uninhabited. However, such events can produce fires within at least 1000 m radius, which could be important in this forested area if they were not controlled.

In contrast, tephra-fall represents a potential hazard if the wind is blowing from the south to north. In this situation, ash could accumulate on the roofs of the villages nearby, with 10 mm of ash adding 19 kg to each m^2 of roof (Alexander 1993). It is also possible that fires could begin in the surrounding area if the ash retained its heat. The possible loss of crops or forest could cause substantial environmental damage.

The 1706 eruption did not pose a hazard in terms of volcanic gases. Even so, there have been two occasions when poisonous gases have been generated during Canarian volcanic eruptions (1582 and 1730–1736): the first produced human victims and both killed animals. The gases can also contaminate spring waters and soils; as there are no rivers on the island, water is supplied by galleries spread throughout the island and the possibility of contamination of the groundwater in the area is therefore another cause for concern.

Conclusions

In general, it is concluded that with Canarian effusive eruptions there is enough time to evacuate the population from the region. Even though the major danger appears at the first moment of the eruption, when the main length of the flow develops (Kilburn 1993), important hazards may also develop from rupture of the lava channel sides and by the development of new active flows, as in the case of the second stage of the eruption at Garachico. These events can be prevented, or the effects mitigated, if the hazard is carefully monitored. However, it is essential that adequate planning by the authorities is undertaken relating to pre-, sin- and post-eruptive measures. It is also important that the population is educated, so that they can respond adequately to these plans.

This study was partially funded by the project 'Teide European Laboratory Volcano' ref. CE: EV5V- CT93- 0283. We especially thank V. Araña, C. Kilburn, A. Hernández-Pacheco and R. Ortiz for their support and advice, and M. C. García, L. García and I. Solana for their logistic help.

References

ALEXANDER, D. 1993. *Natural Disasters*. University College London Press.
ARAÑA, V. 1989. Evaluación del riesgo volcánico en la zona de Santiago del Teide. *Proceedings of the E.S.F. Meeting on Canarian Volcanism*, 260–266.
―― 1991. Canarian volcanism. *Cashiers du Centre Européen de Géodinamique et de Séismologie*, **4**, 13–24.
―― & ORTIZ, R. 1993. Riesgo volcánico. In: ARAÑA, V. & MARTI, J. (eds) *La Volcanologia Actual*. Consejo Superior de Investigaciones Cientificas, 277–385.
AYALA-CARCEDO, F. J. 1993. Estrategias para la reducción de desastres naturales. *Investigación y Ciencia*; **Mayo 1993**, 6–13.
BOOTH, B. 1979. Assessing volcanic risk. *Journal of the Geological Society of London*, **136**, 331–340.

CARRACEDO, J. C., SOLER, V., RODRIGUEZ-BADIOLA, E. & HOYOS, M. 1990. Zonificación del riesgo para erupciones de baja magnitud en la Isla de Tenerife. *Proceedings of the V Reunión Nacional de geología ambiental y ordenación del territorio*, Gijón, Septiembre 1990, 65–72.

CHESTER, D. K. 1993. *Volcanoes and Society*. Edward Arnold.

KILBURN, C. R. J. 1993. *Lava crusts, a flow lengthening and the pahoehoe-aa transition. In:* KILBURN, C. R. J. & LUONGO, G. (eds) *Active lavas: Monitoring and Modelling*. University College London Press, 263–280.

PONTE-COLOGÁN, A. 1911. *Memoria Histórico Descriptiva de esta Erupción Volcánica Acaecida el 18 de Noviembre de 1909*. Tripolit, Tenerife.

ROMERO, C. 1991. *La manifestaciones volcánicas históricas del Archipiélago Canario*. pHD Thesis, Universidad de La Laguna.

ROMERO, J. & BONELLI, J. M. 1951. *La Erupción del Nambroque Junio–Agosto (1949)*. Talleres del Inst. Geograf. y Catastral, Madrid.

SANSÓN, J. 1996. La protección civil ante el riesgo de erupciones volcánicas. *In:* ORTIZ, R. (ed.) *Riesgo Volcánico*. Serie Casa de los Volcanes, **5**, 197–216.

SOLANA, M. C. 1998. *Evaluación de la peligrosidad volcánica en Tenerife a partir de la reconstrucción de cuatro erupciones históricas*. PhD Thesis, Universidad Complutense de Madrid.

UNESCO, 1972. *Report of consultative meeting of experts on the statistical study of natural hazards and their consequences*. Document SC/WS/500.

Index

Page numbers in *italics* refer to Figures and page numbers in **bold** refer to Tables

acid gases
  critical loading 116
  Hekla output 114–115
  sulphur dioxide 167–169
acid rain, volcanogenic 34, 41
Açores, Furnas volcano study
  environmental impacts 198–201
  eruption history 194–198
  socio-economic impacts 203–204
Aegean Arc, Nisyros study 69–71
  EDS analysis
    methods 78–79
    results 79–84
    results discussed 84–86
  pumice fall isopachs 77
  stratigraphy 71–77
  tephrostratigraphy 77–78
aeromagnetic data, Auckland volcanic field 4, *5*, 8–9
aerosols 167
  see also acid gases
Akira Pumice Formation *50*, 51
albedo, volcanic effect on 167
alkali basalt 3
Angat Kitet Formation 51, *52*
anhysteretic remanent magnetization **131**
archaeological applications of tephra
  Bronze Age volcanism 110, 119–120
Ash Hill *2*, **6**
Auckland volcanic field 2–4, *29*
  eruption timing 7–8
  geophysical data 4
  structural style 4–7
  volcanic hazards 29–30, 33–34

back-field ratio **131**, 132
basalt 3, 212
basanite 3, 212
bog oak, impact of Hekla on 118–119
British tephra sites 156–157
Bronze Age
  Hekla eruption 119–120
  Scotland 110

^{14}C, Kaharoa tephra 16

Canary Islands 209
  Negra Mt study
    1706 eruption 210–212
    risk assessments 212–215
charcoal 20
chemistry *see* geochemistry
Chinyero 209
Chironomidae 111
climate change
  effect of Hekla 110–111, 114–115, 118–120
  effect of Laki 112, 113, 161–162
    scientific measures 164–165
    sulphur dioxide effect 166–169
    written accounts 162–165
  effect of Pinatubo 113
  effect of Tambora 112, 113–114
Coromandel Volcanic Zone 28
Corylus pollen record 111
Crater Hill *2*, 4–5, **6**, 7, 8, 9
critical load concept 116
Curie temperature **131**, 132

discriminant function analysis (DFA) 149
  British tephra sites 156–157
  data handling 149–150
  Icelandic tephras 150–153, 154–156

earthquakes *see* seismicity
Egmont, Mt 11, *12*, 28, 32–33
Emuruangogolak 63
energy dispersive spectrometry (EDS)
  methods 78–79
  results 79–84
  results discussed 84–86
Enkorika Formation 52
Esinoni Formation 51, *52*
Etna, 1977–1991 survey
  eruptive phases 96–102
  magnitude/frequency analysis 102–105
  results discussed 105–106
  seismic events 90–96

fire fountains 3
frequency dependent susceptibility **131**

INDEX

Furnas volcano
  environmental impacts 198–201
  eruption history 194–198
  socio-economic impacts 203–204

gases *see* acid gases
geochemistry of tephra 52–55
  DFA study 150–153
    sodium problem 153–154
  eruption correlation 55–61
  Nisyros tephra **80, 81**
geomagnetism *see* magnetic analysis
geophysics 4, *5*, 8–9
Greece, Nisyros study 69–71
  EDS analysis
    methods 78–79
    results 79–84
    results discussed 84–86
  pumice fall isopachs 77
  stratigraphy 71–77
  tephrostratigraphy 77–78
Green Hill *2*, **6**, 8
Gregory Rift, Longonot/Suswa study 47, *48*
  caldera collapse 49–52
  tectonic setting 61–65
  tephrochronology 55–61
    methods 52–54
    results 54–55
Grímsvötn tephra **150**, 150–153

Hampton Park *2*, **6**, 7, 8
Haroharo 28
hazard assessment and management
  Auckland volcanic field 3–4, 29–30, 33–34
  Açores 194–204
    dominant approach 190–192
    radial approach 192–194
Hekla
  tephra analysis by DFA **150**, 150–153, 154–156
  tephra impacts 110–111, 114–115
    Ireland 118–119
    Scotland 119–120
high field strength elements 54–55

Iceland
  Hekla
    tephra analysis by DFA **150**, 150–153, 154–156
    tephra impacts 110–111, 114–115
      Ireland 118–119
      Scotland 119–120

Laki fissure volcano
  eruption dynamics 174
  impact in Europe 111, 175, 176
    air quality 162–163, 177–179
    climate change links 112, 113, 161–162, 166–169
    social responses 183–185
    storminess *176*, 180–183
  impact in Iceland 34, 174–175
Storalda moraine complex
  stratigraphy 126–130
  tephra magnetic analysis
    methods 130–132
    results 132–140
    results discussed 140–144
ignimbrite 51
Indonesia 193
  Tambora 112, 113–114
Ireland
  bog oak case study 118–119
  tephra sites 156–157
isothermal remanent magnetization **131**
Italy, volcanic emergency planning 193
  Etna 1977–1991 survey
    eruptive phases 96–102
    magnitude/frequency analysis 102–105
    results discussed 105–106
    seismic events 90–96

Kaharoa tephra *12*, 36
  analysis **15**
  ^{14}C dating **16**
  environmental impact 17–21, 21–23
  stratigraphy 15
Kaipo bog *28*, 32
Kapenga 28
Kawakawa tephra 36
Kedong Valley Tuff Formation *50*, 51
Kenya Rift Valley, Longonot/Suswa study 47, *48*
  caldera collapse 49–52
  tectonic setting 61–65
  tephrochronology 55–61
    methods 52–54
    results 54–55
Kildonan, Strath of 111
Kilombe 63
Kohoura *2*, **6**
Kohuora Crater *39*, 40
Kos Plateau Tuff 69

Laki fissure volcano

eruption dynamics 174
impact in Europe 111, 175, 176
　air quality 162–163, 177–179
　climate change links 112, 113, 161–162, 166–169
　social responses 183–185
　storminess *176*, 180–183
impact in Iceland 34, 174–175
lapilli pumice 74, 75, 76
lava flow hazard management **191**
Longonot 47, *48*
　caldera collapse 49–51, 52
　tectonic setting 61–65
　tephrochronology
　　eruption correlation 55–61
　　methods 52–54
　　results 54–55

McLaughlins Hill *2*, **6**, 8
McLennan Hills *2*, **6**, 8
magnetic analysis
　basalts 7–8
　tephra
　　methods 130–132
　　results 132–140
　　results discussed 140–144
magnetic susceptibility **131**
major element analysis **80**, **81**
　see also geochemistry
Mamaku tephra 36
Mangakino *28*
Mangere, Mt *2*, **6**, 8
Mangere *2*, **6**, *37*, 38, 40
Maroa 11, *12*, *28*
Matakana Island 13–15
　pollen analysis 17–21
　stratigraphy 15–17
　tephra impact 21–23
Maungarei tephra 40
Maungataketake *2*, **6**
Mayor Island 11
Menengai 63

natural hazards *see* hazard assessment and management
Nb geochemistry 54–55
Negra Mt
　1706 eruption 210–212
　risk assessments 212–215
nephelinite 3
New Zealand
　Auckland volcanic field 2–4, *29*

eruption timing 7–8
geophysical data 4
structural style 4–7
volcanic hazards 29–30, 33–34
Matakana Island 13–15
　pollen analysis 17–21
　stratigraphy 15–17
　tephra impact 21–23
North Island
　tephra depth survey 36
　volcanic centres *28*, 30–33
Ngauruhoe, Mt *28*, 32
Nikia Rhyolite 69
Nisyros 69–71
　EDS analysis
　　methods 78–79
　　results 79–84
　　results discussed 84–86
　pumice fall isopachs 77
　pyroclastic stratigraphy 71–77
　tephrostratigraphy 77–78

Okataina Volcanic Centre 11, *12*, 15, *28*, 31
Olgumi Formation 51, *52*, 54
Olongonot Volcanic Formation 51
Oloolwa Formation 51, *52*
Omapere, Lake 36
Otara Hill *2*, **6**, 7, 8
Otuataua *2*, **6**

palynology 17–21, 111
Patiki Road section 38, 40
phonolite 212
phreatomagmatism 51
　Auckland 3
Pinatubo, Mt 113–114, 167
Pinus sylvestris pollen record 111
pollen analysis 17–21, 111
Pukaki Crater *2*, 4, 5, **6**, 7
Pukeiti *2*, **6**
Pukekiwiriki *2*, **6**, 7
Puketutu *2*, **6**, 7, 8, 9
pumice 51

radiocarbon dating **16**
Rangitoto 3
Reporoa *28*
Richmond, Mt *2*, **6**
risk assessment 212–215
Robertson Hill *2*, **6**, 7
Rotoehu Ash 36
Rotorua *28*

Rototuna, Lake 36
Ruapehu, Mt *28*, 32, 40

Saksunarvatn tephra **150**, 153
San Juan 209
São Miguel, Furnas volcano study
　environmental impacts 198–201
　eruption history 194–198
　socio-economic impacts 203–204
saturation isothermal remanent magnetization **131**
Scotland, Bronze Age study 110
　impact of Hekla 119–120
seismicity and volcanic eruption, Etna 90–96
Sicily, Etna 1977–1991 survey
　eruptive phases 96–102
　magnitude/frequency analysis 102–105
　results discussed 105–106
　seismic events 90–96
Silali 63
Skaelingisvatn 111
sodium problem in geochemical analysis 153–154
statistical techniques in DFA 149–150
Storalda moraine complex
　stratigraphy 126–130
　tephra magnetic analysis
　　methods 130–132
　　results 132–140
　　results discussed 140–144
Styaks Swamp 2, **6**, 8
sulphur dioxide 167–169
Suswa 47, *48*
　caldera collapse 51–52
　tectonic setting 61–65
　tephrochronology
　　eruption correlation 55–61
　　methods 52–54
　　results 54–55
Svinavatn tephra **150**, 150–153

Tambora (Indonesia) 112, 113–114
Tarawera *12*, 15, *28*
Taupo 11, *12*, *28*, 31–32, 36
temperature *see* climate change
Teneguia 209
Tenerife 209
　Negra Mt
　　1706 eruption 210–212
　　risk assessments 212–215
tephra 125

climatic impact 110, 118–120, 175, 176
　air quality 162–163, 177–179
　climate change links 112, 113, 161–162, 166–169
　social responses 183–185
　storminess *176*, 180–183
discriminant function analysis (DFA) 149–150
　British tephra sites 156–157
　Icelandic tephras 150–153, 154–156
energy dispersive spectrometry (EDS)
　methods 78–79
　results 79–84
　results discussed 84–86
geochemistry 52–55
　eruption correlation 55–61
magnetic analysis
　methods 130–132
　results 132–140
　results discussed 140–144
tephrochronology 125
　Kenya Rift 55–61
　methods 52–54
　results 54–55
　New Zealand **16**, 21–23, 36
tholeiite 3
Timanfaya 209, **210**
Tongariro 11, *12*, *28*, 32
trace elements analysis **59**, **60**
　see also geochemistry
trachybasaltic ash 51
trachyte agglutinate 51
trachyte globule ignimbrite 51
trachyte pumice lapilli 51
tsunami 30
Tuhua 11, *12*, *28*, 31
Tutira, Lake 32

volcanic gases *see* acid gases
vulnerability analysis 194

Waiatarua, Lake 39–40
Waikato lakes *28*, 36
Waitomokia 2, **6**, 7
weather *see* climate change
Wellington, Mt 40
Whakamaru *28*
Wiri 2, **6**, 7, 8, 9

Zr geochemistry 54–55